Smart Manufacturing - When Artificial Intelligence Meets the Internet of Things

Edited by Tan Yen Kheng

Published in London, United Kingdom

IntechOpen

Supporting open minds since 2005

Smart Manufacturing - When Artificial Intelligence Meets the Internet of Things
http://dx.doi.org/10.5772/intechopen.90084
Edited by Tan Yen Kheng

Contributors
Amandeep Sharma, Pawndeep Sharma, Yinquan Yu, Christopher Toh, James Brody, Abhra Roy Chowdhury, Subin Raj, Prajwal Prabhu, Malcolm H. Smith, Tan Yen Kheng, Felix George

First published in London, United Kingdom, 2021 by IntechOpen
IntechOpen is the global imprint of INTECHOPEN LIMITED, registered in England and Wales, registration number: 11086078, 5 Princes Gate Court, London, SW7 2QJ, United Kingdom
Printed in Croatia

British Library Cataloguing-in-Publication Data
A catalogue record for this book is available from the British Library

Additional hard and PDF copies can be obtained from orders@intechopen.com

Smart Manufacturing - When Artificial Intelligence Meets the Internet of Things
Edited by Tan Yen Kheng
p. cm.
Print ISBN 978-1-83881-087-0
Online ISBN 978-1-83962-649-4
eBook (PDF) ISBN 978-1-83962-650-0

An electronic version of this book is freely available, thanks to the support of libraries working with Knowledge Unlatched. KU is a collaborative initiative designed to make high quality books Open Access for the public good. More information about the initiative and links to the Open Access version can be found at www.knowledgeunlatched.org

For EU product safety concerns:
IN TECH d.o.o., Prolaz Marije Krucifikse Kozulić 3, 51000 Rijeka, Croatia, info@intechopen.com or visit our website at intechopen.com.

Meet the editor

Dr. Tan Yen Kheng is the CEO and co-founder of Printed Power, a high-tech company headquartered in Singapore that develops edge computers for smarter buildings and manufacturing globally (China and the Association of Southeast Asian Nations [ASEAN]). The company empowers customers with sensor-end to application-end platform solutions to discover opportunities and capture value from actionable insights as well as co-create with domain experts using advanced artificial intelligence (AI)/machine learning (ML) tools. Dr. Tan is concurrently the associate editor of the IEEE Sensors Journal and the industrial chair of IEEE Singapore section. He was also chair/professor at the School of Electrical Engineering/Hanergy School of Renewable Energy at Beijing Jiaotong University (BJTU) where he built international partnership programs, delivered seminars and short courses, and performed research exchanges with research staff and students.

Contents

Preface XI

Section 1
Smart Manufacturing 1

Chapter 1 3
Remote Management of Autonomous Factory
by Yen Kheng Tan and Felix George

Chapter 2 15
Optimization of Manufacturing Production and Process
by YinQuan Yu

Chapter 3 49
Interlinking Industry 4.0 and Academia through Robotics and Automation: An Indian Perspective
by Prajwal Prabhu, Subin Raj and Abhra Roy Chowdhury

Section 2
Artificial Intelligence-of-Things (AIoTs) 63

Chapter 4 65
Applications of Machine Learning in Healthcare
by Christopher Toh and James P. Brody

Chapter 5 91
Radio Systems and Computing at the Edge for IoT Sensor Nodes
by Malcolm H. Smith

Chapter 6 131
Energy Harvesting Technology for IoT Edge Applications
by Amandeep Sharma and Pawandeep Sharma

Preface

Smart Manufacturing for tomorrow will develop a generic assembly-centric production shopfloor based on internet-of-things (IoT) and internet-of-services (IoS) technologies served by a high-value workforce with a key focus on dynamical reconfigurability of production to cope with ever-increasing demand in low-volume high-mix production.

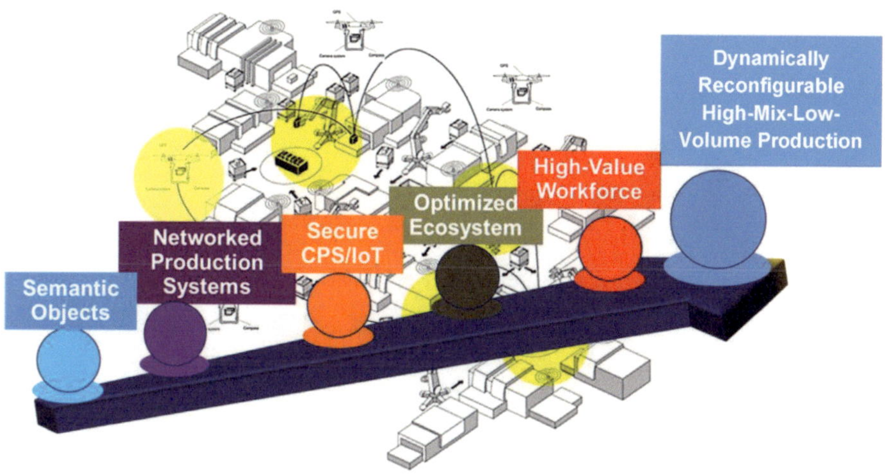

Around the world there are a few notable smart manufacturing initiatives paving the way to the future. Some of these initiatives include Made in China 2025 (MIC2025), "Industrie 4.0" (I4.0) in Germany, the Industrial Internet (Made in America [MIA]), the Smart Nation in Singapore, and the Factory of the Future in the United Kingdom/European Union. Clearly, the global trend is pivoting towards a new era, away from the traditional mass factory production pioneered by Henry Ford to a consumer-oriented low-volume, high-mix demand and supply production. The consensus worldwide is that the next generation of manufacturing will involve the transformational convergence of intelligent machines, powerful computing and analytics, and unprecedented networking of people, products, and services. Various visions to this end (e.g., "Industrie 4.0," MIA, and MIC2025) share this sentiment and are being pursued by companies as they realize the tremendous potential payoffs, including optimization of assets, facilities, and networks as well as new business opportunities that have been estimated to add $10-$15T to the global GDP (Evans & Annunziata, 2012).

Big data has revolutionized to be the heart of everything. Suppliers, manufacturers, consumers, and backend logistic supply chains are already interconnected among themselves by the Internet of things (IoT) and so, naturally, data is being packaged into the Internet of services (IoS) to serve the smart manufacturing of tomorrow. Manufacturing-as-a-service (MaaS) will drive smart manufacturing companies towards dynamic and reconfigurable production through sharing the costs of equipment, software, maintenance, and repairs. This setup creates a sharing economy between businesses and reduces the manufacturing costs for each involved, allowing companies to expand more quickly and efficiently because they are not forced to pay for the expansion on their own. Furthermore, personalization is the new frontier of consumer goods; delivering a personalized experience is key in the age of the hyper-connected consumer. The Internet of People (IoP) will fulfill and embrace each individual's need with big data at the heart of everything.

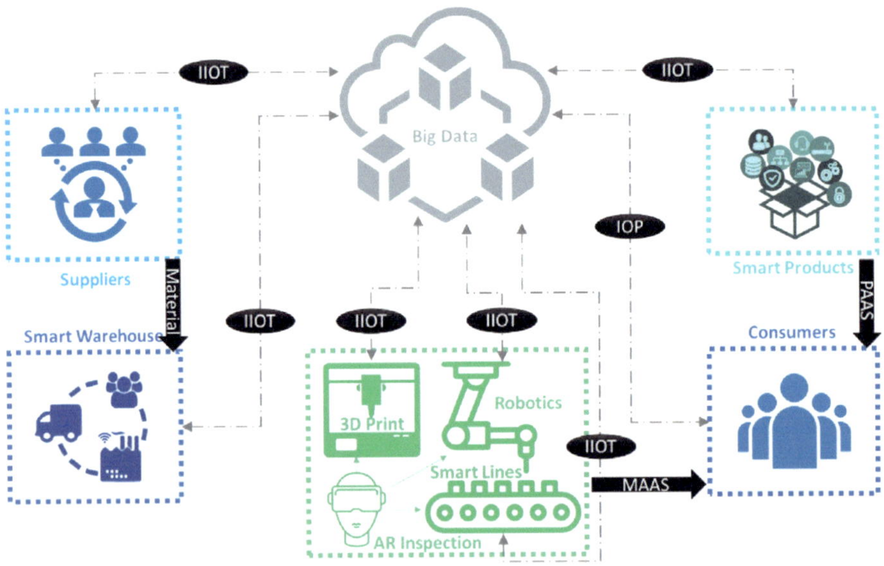

Within the ongoing smart manufacturing transformations, there are a series of developments on increasing-fidelity models from a manufacturing facility to individual machines as appropriate and developing strategies for improving operations or creating new business opportunities. The relevant emerging trends include:

- New technologies involving increased Internet adoption, e.g., IoT, cloud computing and services, and mobile platforms with unprecedented multimode sensors and global connectivity.

- New and unexpected markets enabled by paradigms such as Design for X, social networks, the maker movement, Novel Architecture, and the IoT and IoS.

- New business boundaries resulting from the collapse of existing supply chains and the emergence of novel value networks where value chain positioning may be restructured and firm boundaries become blurred.

- Uncertainty and risk of selecting new technology and applications. Semiconductors, microelectromechanical systems (MEMS), and multimode sensor technologies of mobile communication devices and robotics are evolving, and various companies are battling for dominant positions in newly created arenas.

Hence, it is essential to explore the usefulness of these digitalization tools for smart manufacturing applications with a focus on the convergence of IoT, artificial intelligence (AI), and big data into the edge computing era. Cloud computing is no longer the ideal solution to support 5G connectivity without limiting the power to automate, to think emotionally, and to uplift the various sectors in large-scale development. This book introduces and discusses various technologies in the smart manufacturing arena where AI meets with the IoT to create explosive impacts.

Yen Kheng Tan Ph.D.
Research Curator,
Printed Power Pvt. Ltd.,
Singapore

Section 1

Smart Manufacturing

Chapter 1

Remote Management of Autonomous Factory

Yen Kheng Tan and Felix George

Abstract

In today's mass production era, the world is making things (products and systems) so quickly and systematically in huge volume. The demand for these products is very high and, at the same time, consumers are still in search for a need for making the production very personalized. Hence, the "one mold fits all" approach may not seem to be enough. The present approach is facing the lack of networking between the automation pyramid levels, that is, especially between enterprise resource planning (ERP) and manufacturing execution system (MES) layers and, in turn, communicating directly with the lower layers is not possible. This missing communication among the process equipment like machineries and field control systems like PLCs at the production shop floors implies that customization at the product layer for the consumer is still in progress in classical manufacturing. Mini-MES is a new concept being introduced here to solve the existing techniques reported in the literature and is followed by industry best practices. The novel mini-MES platform provides an avenue for the technology process level (the most bottom layer) to interplay interconnectivity and interoperability with its higher levels until the above pain points are addressed holistically. The chapter is going to focus mainly on the factory production of digital manufacturing and on describing the 3-Cs implementation plan, the enabling technology, and the achievable outcome ahead.

Keywords: factory cloud platform (FCP), autonomous factory, edge computing AI

1. Introduction

The smart manufacturing initiative from the industry transformation perspective is rapidly emerging in many parts of the world. Few notable smart manufacturing transformations leading the globe are the Chinese version of Made in China 2025, the German version of Industry 4.0, the United States version of Industrial Internet (Made in America), the Singaporean Smart Nation, and the UK/EU version of Future of Factory. Clearly, the global trend is pivoting toward the new era from the traditional mass ("Ford's") factory production, where a consumer-oriented low-volume, high-mixed demand and supply production would be the smarter way to build and make things [1].

In the factory, the manufacturing system is illustrated by the pyramid shown in **Figure 1**. At present, the production shop floor is already highly reconfigurable in its process level. This is possible, starting from the enterprise resource planning

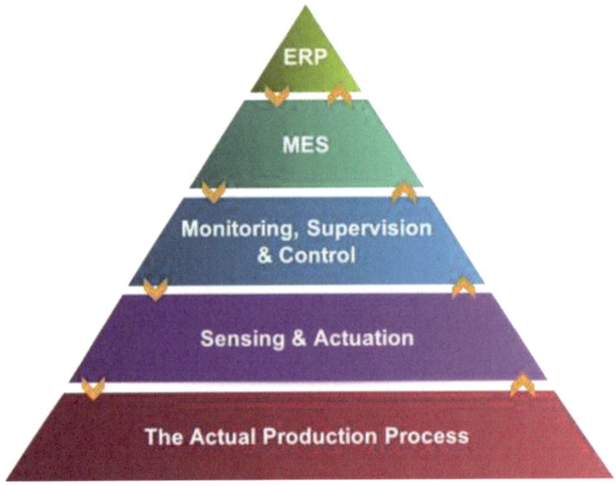

Figure 1.
The manufacturing pyramid—an overview of the existing production flow [2].

(ERP) layer, where process planning, operational management, etc. are interlinked. The manufacturing execution system (MES) layer is set to handle scheduling, dispatching production, etc. in a systematic manner. Monitoring, supervision, and control layer takes care of the real-time optimization and the advanced process controls. Sensing and actuation layer comprises the sensor readings' collection and manipulation of the production process. The final layer is where the actual production process exists.

In the above manufacturing pyramid diagram, few layers are already interconnected, but there are some layers which are independent silos. This is the pain point that is addressed through smart manufacturing, that is, introduction of the smart connectivity and interoperability throughout the layers. Various aspects need to be considered as it could be impossible to replace all the hardware in a manufacturing company to enable this. There could be some legacy systems as well which in itself would not be ideal to use in such an environment. By introducing the right cyber-physical tools and network connectivity, this can be achieved.

2. Digital manufacturing transformation

2.1 Mass production era

In today's mass production era, the world is making things (products and systems) so quickly and systematically in huge volume. The demand for these products is very high and, at the same time, consumers are still in search for a need for making the production very personalized. Hence, the "one mold fits all" approach may not seem to be enough. The present approach is facing the lack of networking between the automation pyramid levels, that is, especially between ERP and MES layers and, in turn, communicating directly with the lower layers is not possible. This missing communication among the process equipment like machineries and field control systems like PLCs at the production shop floors implies that customization at the product layer for the consumer is still in progress in classical manufacturing [3].

Recently, there is a strong pull to shift from the classical manufacturing/automation pyramid to the horizontal and vertical networking and to achieve the cyber-physical production systems (CPPS) as illustrated in **Figure 2**.

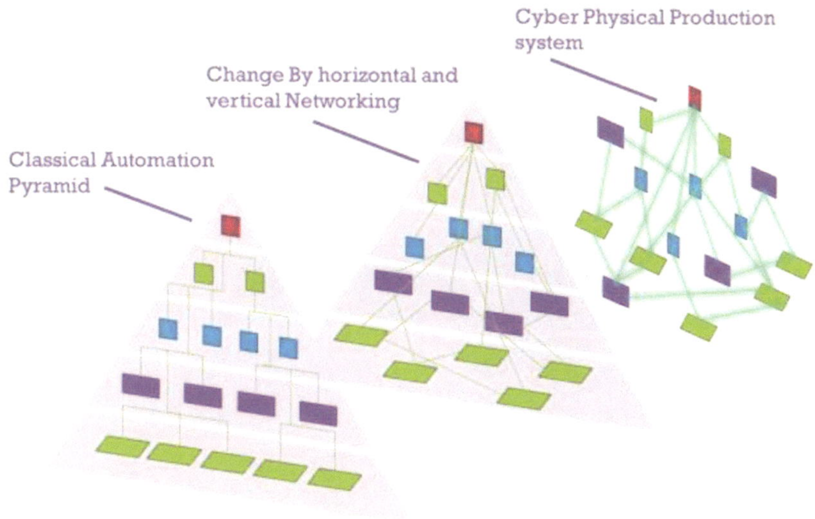

Figure 2.
The new paradigm of the manufacturing pyramid—an overview of the changed production flow.

The hierarchy arrangement of the automation pyramid is helpful for tracking the production status and for keeping proper process flow. Chaining the horizontal levels together adds value to the coordination and the vertical levels networks for more interconnection and cooperation. The full potential is attained when the boundaries are open. To reach this CPPS goal, one has to first identify and determine the associated pain points in order to enable the smooth transition:

1. Vast proprietary silos platforms/systems out there in the industrial systems.

2. Proper utilization of the available data.

3. Demand for higher response in terms of demand and supply of products in between the production line and the supply chain flow.

2.2 Digital manufacturing era

The CPPS is the backbone to the digital manufacturing era. The transition process from classical to digital manufacturing transformation requires a holistic platform of many Mini-MESs as shown in **Figure 3**. Mini-MES is a new concept being introduced by the authors to solve the existing techniques reported in the literature and is followed by industry best practices. The novel mini-MES platform provides an avenue for the technology process level (the most bottom layer) to interplay interconnectivity and interoperability with its higher levels until the above pain points are addressed holistically. There are four main aspects to be looked at and addressed, namely product, production, process management, and service manufacturing. This platform works hand in hand with the two upper layers to form the full CPPS as mentioned before in achieving the autonomous production factory. The rest of the chapter is going to focus mainly on the factory production of digital manufacturing and on describing the 3-Cs implementation plan, the enabling technology, and the achievable outcome ahead.

The 3-Cs implementation plans can be divided into three main parts. First is connecting to enable remote monitoring and management [5]. To truly remotely

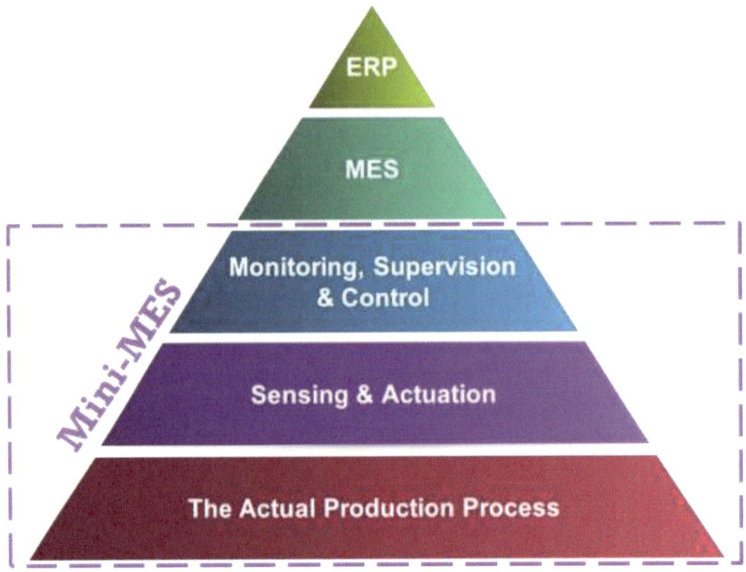

Figure 3.
Paradigm shift from classical automation toward interconnected smart systems [4].

manage the factory in real time, 5G technology has to be used since the specifications of 5G enable us to achieve a near real-time operation. The desired outcome is to achieve the large-scale deployment of remote autonomous control systems that are to be operated and managed seamlessly.

Second is computing at the edge. To implement as an aggregation point of the factory that holistically connects the dots of the individual levels of the automation pyramid and seamlessly control the communication protocols of devices and equipment of Siemens, Schneider Electric, ABB, etc. with a single common protocol. By doing so, any data point on the factory shop floor is available to the A.I. computation to develop the desired models and throughputs.

Third is collaborating/codeveloping with the domain experts. To leverage on the strong client records of experiences and add on to the implementation of the data platform to drive the traditional manufacturing market in a much faster and penetrable manner. Through the codevelopment process, the digital transformation of factory automation is going to accelerate tremendously while working closely back and forth with existing clients and leveraging on the existing manufacturing solution and the domain expertise.

3. Toward autonomous factory

As you may already realize, the global manufacturing trend is changing from the passive and reliant approach to a more demanding and intelligent way to manufacture produces. That is the smart factory (SF), which is rapidly growing and will be an anchor component of the digital manufacturing transformation journey. According to a financial report, it is expected that the SF market will continue to grow at a compound annual growth rate (CAGR) of 9.6% to reach around US$244.8 billion by 2024. In the case of the autonomous factory, it is projected to be one of the key enablers of these SFs for having the ability to be remotely operated and managed in the autonomous manner. Thanks to the data-driven platforms that are booming in from the digital space, the factory production can turn toward full automation and autonomy. It is expected to transform the traditional factory owners that produce in mass quantities with lots of manual workers into low-volume, high-mixture production with digital manufacturing technique.

The transformation will involve a set of digital technologies, including the Internet of Things (IoT), robotic and artificial intelligence (AI), into the implementation/upgrade for the factory modernization. With this, the new world of making things so agile in a flexible way, optimizing and integrating the system approach, and the great speed and accuracy are going to benefit and satisfy the fast consuming and highly personalized needs of the people (**Figure 4**).

Remote monitoring is a key aspect for an autonomous factory. For a manufacturing company to make money, it ultimately comes down to the number of products made. So, if there is an unscheduled downtime, the throughput of the production line reduces, and hence the company ends up making fewer products and thus loses money. We can overcome this by introducing a remote management solution. This would consist of a list of tools such as remote production monitoring, remote safety monitoring, machine learning, and artificial intelligence for predictive maintenance, which would address the problem even before they occur. As the solution is deployed, the owner will be able to see a reduction in operational costs and an increase in the equipment efficiency.

Remote management and monitoring had been available as a technology for the past decade or so, but somehow, the breakout year was never seen by the experts and observers. However, the situation is poised to change in today's scenario, with the new economic trends, transformed man power needs, and the availability of new technologies. During the global financial crisis, a lot of companies turned to lean staffing models. There is a certain lack of design and maintenance talents in many companies because of this. Coupling together with this is the retirement of the baby boomer workforce [6], who are extremely hardworking, which eventually led to a change in the demography of a factory floor. The new staff severely lack the decades of hands-on experience on the aging legacy equipment. They are also put under tremendous pressure to deliver a better throughput and more uptime and hence forcing the companies to remote monitoring and management.

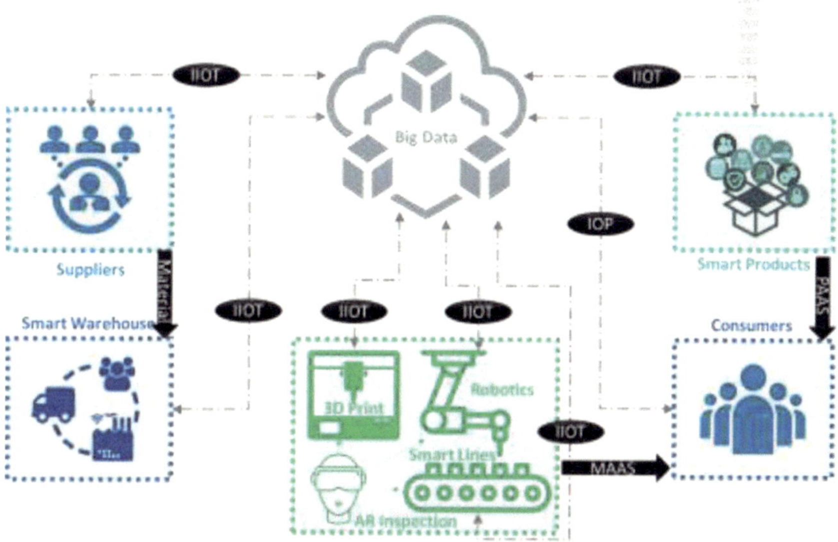

Figure 4.
Autonomous factory flow.

Another challenge we all know well is that the present manufacturing ecosystem is highly dependent on China, where most of the factories are located. China has a very strong fire wall, since they need to ensure their cyberspace is well protected, which is understandable. There is a great need for an IT platform that is equipped with the ability to capture true value creation from opening up the interrelationship between the nations like Singapore and China. The data exchange channel link has to be interlinked with minimum latency to empower an autonomous factory to interact on a real-time basis with its clients throughout day and night for 365 days without fail and to report machine health, process flow rate, and production status to the health diagnostic center for close and constant tracking outside China. Take for an example, there is this possibility with the use of the newly built Singapore and China data connectivity infrastructure of less than 60 ms latency between countries to enable online live communicating and producing of productions. This powerful data transmission exchange infrastructure will definitely enable an autonomous factory to work in a well-protected network environment.

The way things are bought and sold is going to change. It will go, way beyond buying from ecommerce platforms to making the products directly for the consumers. Take for an example, a production factory based somewhere in the middle part of China, in a city called Henan province, where there are a large varieties of machining/processing tools on the actual manufacturing shop floor. The digital manufacturing technology made by Printed Power (PP) Pvt. Ltd. (a company based in Singapore focusing on IoT and smart manufacturing) for the plant is an IoT-cloud platform that allows the remote manning, from the other parts of China or even from Singapore, of the Henan's factory production and maintenance as well as interconnecting the local resource planning ERP with the manufacturing execution system. The power of mini-MES (as illustrated in **Figure 3**) autonomous factory platform is to do away with: (1) restricted remote connectivity for the manufacturing solution to provide access to their local factory floor within China and (2) limited 24 × 7 real-time domain experts' monitoring and resource planning and managing of the factory production system.

In short, this proposed mini-MES autonomous factory platform is one of the first types of deployment in the world that compasses IoT-cloud cum edge computing technology to upskill the local manufacturing solution providers in serving their end manufacturing factory clients more effectively and efficiently. The autonomous factory software platform is performing three key functions: (1) to allow the manufacturing solution provider/system integrator to upgrade the traditional factory to be remotely access and control and (2) to support the manufacturing process/operation with the at-the-edge computers to do data compression for fast speed production and for onsite decision and execution for real-time factory throughput.

4. Technology enablers for autonomous factory

In this section, we will discuss the key technology enablers of an autonomous factory system. The discussion is going to surround on the digitalization tools, namely, the Internet of Things (IoT) hardware, the AI machine learning software, and the application design thinking/innovation needs. **Figure 5** illustrates an overview of the various technology enabling modules that apply to the autonomous factory context.

Among the many, the middleware or the firmware and operating system layers are the most important for the distributed edge computer network with AI processing at the cloud edge. As so, the edge computer becomes a crucial

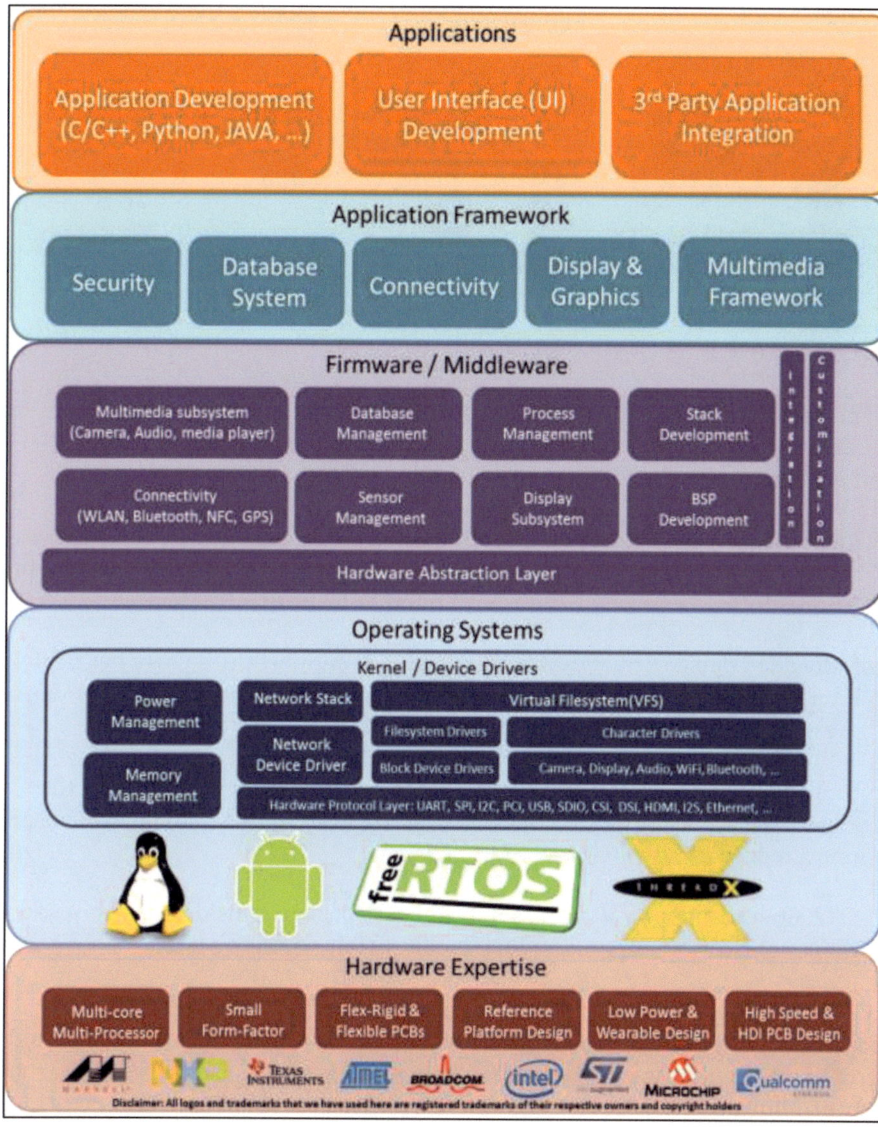

Figure 5.
Technology enablers (reference to https://volansys.com/product-engineering-services/).

technological enabler to the transformation of mass manufacturing into low-volume, high-mixed production. The discussion will focus on the product and system of edge computing and its underlying features as well as its association with manufacturing application and solution. As a reference, some of the tools used by Printed Power will be discussed.

4.1 Printed Power PowerEdge™

Printed Power has developed an IoT edge gateway product as shown in **Figure 6**, which is a powerful hardware platform with multiple protocol support for wired and wireless communication systems. Edge gateway product supports Android, Linux, and Ubuntu operating systems out of the box and can run a multitude of "apps" for each platform. It is a compact industrial PC combined with a powerful IoT gateway.

Figure 6.
Printed power PowerEdge™ hardware specification.

The PowerEdge™ product consists of RK3399 SOC, wireless radios, power modules, multiple GPIO and USB ports, and camera ports enclosed in a weather-proof enclosure with the following:

a. Light-weight operating system (OS): quick booting time and selected services running to reduce the dynamic memory consumption. The OS is based on the trimmed-down version of Android, Linux, or Ubuntu.

b. Embedded device driver development: able to support third party IoT devices interfaced over wired or wireless connectivity. Device drivers are dependent on the current OS running on the platform.

c. Embedded hardware development: able to install add-on hardware for the edge gateway to support IoT devices which operate at different power levels or nonstandard protocol.

d. OTA update support: allow updatable library packages inside the edge gateway for easier firmware upgrade which is completely wireless based.

On top of that, the Printed Power edge-cloud platform depicted in **Figure 7** allows the edge devices that are distributed throughout the factory shop floor and interconnected with the various machineries to gather at one place for monitoring and controlling. The cloud runs machine learning algorithm, makes java class image for download into the edge devices, and interoperates between the cyber cloud and the physical factory seamlessly.

4.2 Digital manufacturing application and solution

Let us first take a look at what a manufacturing production floor looks like and the kind of stuffs they do. A factory, manufacturing plant, or a production plant is an industrial site usually consisting of buildings and machinery, or more commonly a complex having several buildings, where workers manufacture goods or operate machines processing one product into another [4]. The area in a manufacturing facility is where the assembly or production is carried out either by an automated system or by workers or by a combination of both. The shop floor may include equipment, inventory, and storage areas. The basis of manufacturing [7] can be described in the following areas:

1. Assembly line: defines the progression and synchronization of work. The Ford example is typical of manufacturing, where there is a specific sequence of events which must be followed in order to produce the desired work product in a timely manner.

2. Division of labor: breaks the production process into separate tasks performed by specialists or craftsmen. Subdividing the process down into smaller increments provides the means to employ common workers as opposed to developing a dependence on highly skilled craftsmen which may add to the cost to the work product.

3. Precision tooling: provides mechanical leverage in the assembly line, the need for using the most technologically advanced tools, something that requires constant monitoring and upgrading.

Machines and the people are both working together in the shop floor and they operate with each other simultaneously. The machines are controlled by PLC logic and humans are controlled by themselves or by the company process. It is thus important that the two systems of their own talk with each other. **Figure 8(i)** is a snapshot of how this has being designed for an autonomous factory and it illustrates the architecture of such a platform design.

Referring to **Figure 8(ii)**, the robotic automation system of the autonomous factory consists of the arm performing pick and place, the moving robot to convey materials and products inside the factory, and the pressing and heating machine to treat and mold the made products. In order for the machines to communicate by themselves within the horizontal level of the manufacturing pyramid as discussed before, one has to first make a common communication language translator that allows cross talk among the machine systems. Printed Power PowerEdge™ fits in to provide the hardware connectivity as well as the firmware language translation.

Next, some form of intelligence is gathered from the data sets collected from the operations and maintenance to derive with the AI learning models for the control and automating of the manufacturing process. The AI learning is supported by both the online cloud-based engine as well as the offline edge Printed Power PowerEdge™ executable controls. GPU is part of this ball game to power up the AI engine and speed up the entire manufacturing process.

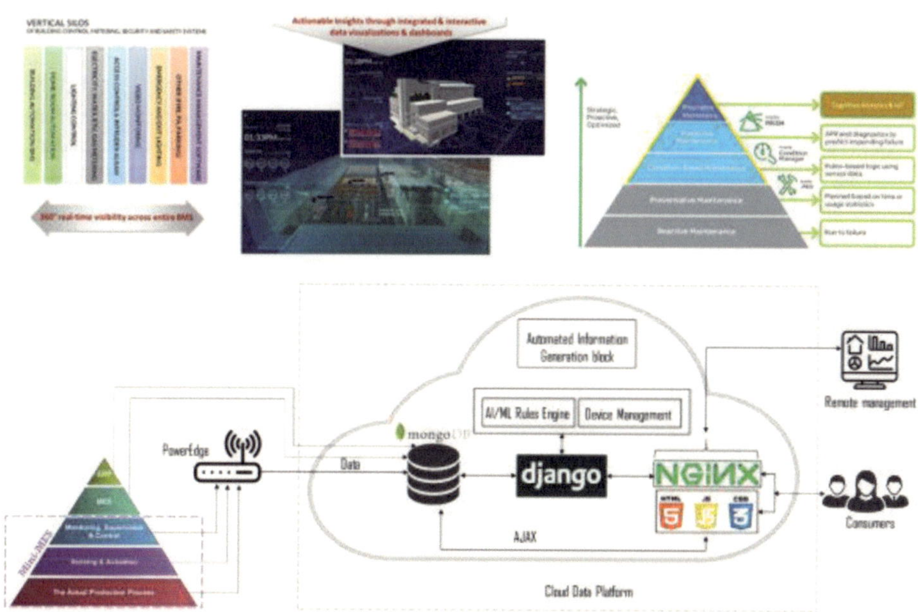

Figure 7.
Printed Power edge-cloud platform. (i) Description of platform components. (ii) Application of edge-cloud.

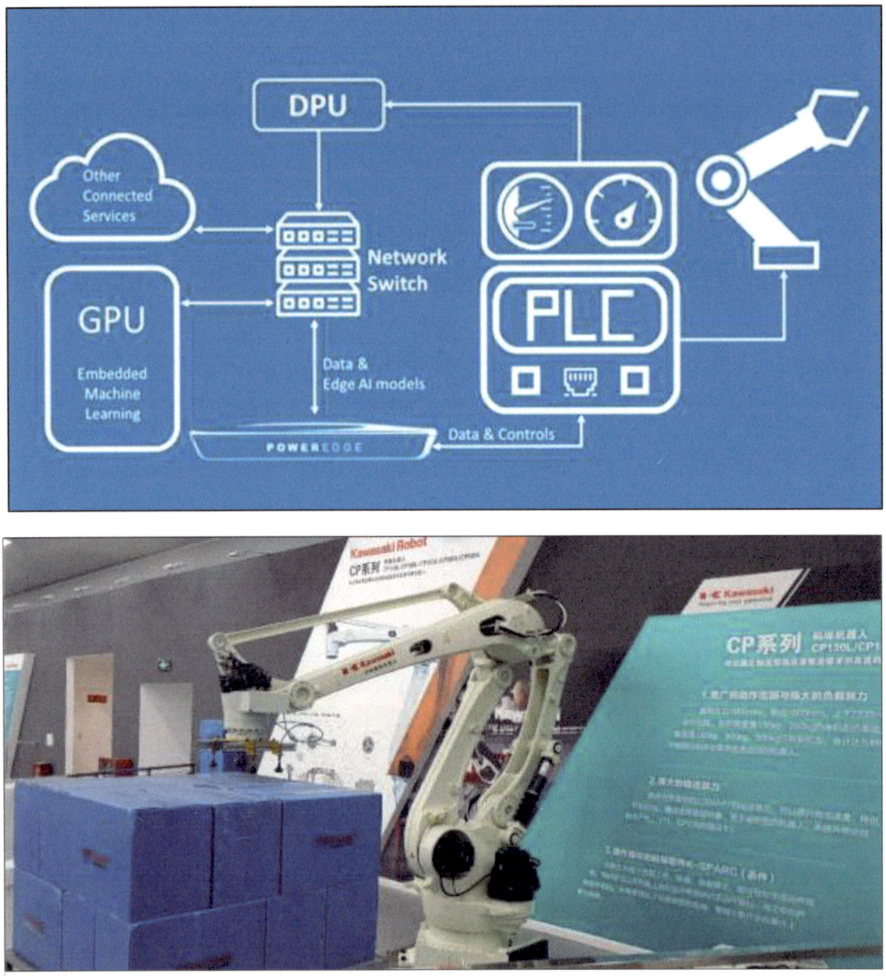

Figure 8.
Autonomous factory platform. (i) Printed Power digital manufacturing architecture. (ii) Robotic automation system.

Lastly, the remote manning and controlling of the entire autonomous factory platform is achieved via the dashboard at the command and control center. With the manufacturing and production status, the business owner is able to make a holistic and accurate decision with the sales figures, logistic and supply chain availability, demand and supply status, etc. Hence, this is a crucial part of the manufacturing application and solution.

The entire platform is undergoing development and installation at the site. It is expected to deliver its desired outcomes of being able to remotely access and control via the designed mini-MES platform with its backend housed on the AWS cloud. The autonomous factory should operate at its highest efficiency with the installed PowerEdge™ edge data compression for fast speed production and for onsite decision and execution for the real-time factory throughput.

5. Conclusion

Smart manufacturing initiative or movement from the industry transformation perspective is rapidly emerging in many parts of the world. Today, mass production is making things (products and systems) in a gigantic way that everyone is consuming them not at their preferences. The made from the "one mold fits all" approach does not seem to work for the consumers. The new world of making things is going to be so agile, so optimized and integrated, and so fast in speed and precise in accuracy to benefit and satisfy the fast consuming and highly personalized needs of the people.

To address this, the chapter discusses the digital manufacturing technology made by a Singaporean company, Printed Power, for the plant to be remotely manned and controlled by a press of a button. This autonomous factory software platform which is being developed allows for: (1) the remote connectivity for the manufacturing solution to access their local factory floor within China and (2) the 24 × 7 real-time outside China monitoring and managing of the factory production system.

Author details

Yen Kheng Tan* and Felix George
Printed Power Pvt. Ltd., Singapore

*Address all correspondence to: tanyenkheng@yahoo.com.sg

References

[1] Available from: https://patch.com/florida/palmharbor/bp--the-five-elements-of-mass-production [Accessed: 13 May 2020]

[2] Rojko A. Industry 4.0 Concept: Background and Overview. Available from: https://online-journals.org/index.php/i-jim/article/viewFile/7072/4532 [Accessed: 13 May 2020]

[3] Business Directory. Available from: www.businessdictionary.com › definition › shop-floor [Accessed: 13 May 2020]

[4] Dietmar Theis. Industry 4.0—Chances and Challenges of the Digital Transformation. Available from: http://www.picmet.org/new/conferences/17/handouts/Keynote_Theis.pdf [Accessed: 13 May 2020]

[5] Available from: https://www.motioncontrolonline.org/content-detail.cfm/Motion-Control-News/New-Trends-in-Remote-Monitoring-Enhance-Manufacturing/content_id/2536 [Accessed: 23 July 2020]

[6] Available from: https://www.thebalancecareers.com/baby-boomers-2164681 [Accessed: 23 July 2020]

[7] Printed Power Singapore. Available from: www.printedpower.sg [Accessed: 13 May 2020]

Chapter 2

Optimization of Manufacturing Production and Process

YinQuan Yu

Abstract

This chapter mainly introduces production processing optimization, especially for machining processing optimization on CNC. The sensor collects the original vibration data in time domain and converts them to the main feature vector using signal processing technologies, such as fast Fourier transform (FFT), short-time Fourier transform (STFT), and wavelet packet in the 5G AI edge computing. Subsequently, the main feature will be sent for cloud computing using genetic programming, Space Vector Machine (SVM), etc. to obtain optimization results. The optimization parameters in this work include machining spindle rotation velocity, cutting speed, and cutting depth, while, the result is the optimized main spindle rotation speed range of CNC, which met machining roughness requirements. Finally, the relationship between vibration velocity and machining quality is further studied to optimize the three operational parameters.

Keywords: machining processing optimization, time domain, short time Fourier transform, wavelet packet, genetic programming, deep learning

1. Introduction

The International Federation of Information Processing (IFIP) defines a numerically controlled (also commonly called **CNC**) machine tool as a machine tool equipped with program control. The difference between CNC machine tools and ordinary machine tools lies in the working sequence of NC machine tools: according to the requirements of part processing, CNC language is used to write processing sequences and parameter programs. After the program is analyzed and processed by the CNC device, the execution instructions are sent to the servo system to actuate the motion of the machine tool [1]. Programming controlled movement of the main spindle and worktable completes the processing that does not exist in ordinary machine tools. As shown in **Figure 1**, the mechanical aspects of CNC machine tools mainly include three major parts: the main shaft component [2], followed by the support component, and the conveying mechanism [3]. The main shaft is partially driven by a high-precision stepper motor or servo motor instead of a conventional motor. The transmission part uses a ball lead screw with less resistance and greater rigidity without backlash instead of the traditional lead screw. The CNC system controls the movement of the mechanical structure of the CNC machine tool to complete the processing of parts. The main component is the MCU. The coordination between its functions is an important index for evaluating the CNC machine tool and the CNC system [4].

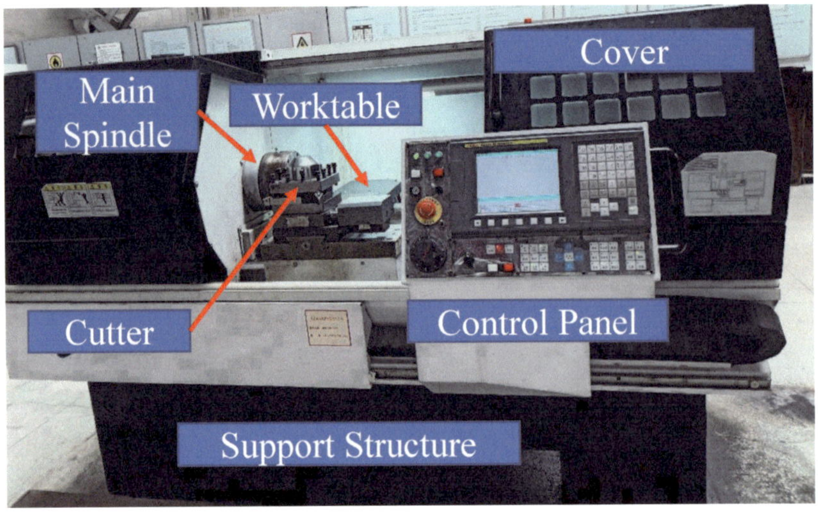

Figure 1.
Components of a CNC lathe.

The optimization of CNC machine tools mainly includes the following three aspects:

1.1 Mechanism optimization

Improving machining accuracy of the machine tool must start from the design stage of the NC machine tool, adopt the drive device with outstanding dynamic performance, and use advanced control technology to improve anti-interference ability of the servo system [5–9]. Structural optimization design can calculate the durability and maintainability of NC machine tools, and design reliability by calculating the space error model of XYZ CNC machine tools by first-order and second-order matrices, Monte Carlo method, etc. [10]. By studying the kinematic configuration of CNC machine tools, it is found that it directly affects the nonlinear errors generated in the process of free-form surface machining [11], and has a direct impact on processing energy consumption and design of the supporting CNC system [12]. The moving parts of CNC machine tools and the mechanism of each feed axis also have a significant impact on the overall rigidity [13], positioning error [14], maintainability [15], and other indicators of CNC machine tools, which determine machining accuracy and work reliability [16]. The spindle of CNC machine tools can be analyzed and optimized by finite element analysis [17]. The finite element modeling can appropriately simplify some chamfers, small holes, etc. that do not affect mechanical properties [18]. The denser and more accurate the finite element (FEM) mesh division, the longer the optimization calculation time [19]. The method of selecting the appropriate unit division, which is not discussed here, can be found in ANASYS-related books. In addition to the influence of the spindle of the CNC machine tool, the deformation of the CNC machine tool due to insufficient static stiffness [20] causes deviations between the actual position and the ideal position of the tool and the workpiece, and seriously affects its machining accuracy. The guide rail has a decisive influence on the accuracy of the machine tool [21]. The optimized design of the guide rail of the machine tool can significantly improve the geometric accuracy of the bed rail, thereby significantly enhancing the machining accuracy of the NC machine tool. Furthermore, selecting the right load rail by speed can optimize the machining quality. In terms of precision machining, a reasonable milling process should be selected according to different materials [22]. Choosing a suitable processing tool and enhancing the strength of the processing tool can avoid vibration caused by the high speed of the machine tool spindle during processing, which will affect processing accuracy [23]. For example, in the process of boring bar technology processing, strong heat treatment technology can be used to improve the rigidity and strength of its material. The error of the tool during manufacturing and machining wear [24] requires us to monitor it in real time to detect problems in time, check and integrate and summarize these problems, and then establish an error compensation model through the data system [25]. The influence of tools and bearings on the accuracy of machine tools is obvious [26]. The machining process should select the tool based on the cutting degree, depth, and accuracy of the parts [27], and the bearing selection should use bearings with relatively low friction resistance and high degree of smoothness and stability. During gear machining, the accuracy of the machine tool will affect transmission characteristics during gear machining [28–30]. The machining process can be regarded as the cutter wheel axis revolving around the center axis of the production wheel. Cogging is processed by 3–7 groups of tools. After each cogging is processed, the workpiece rotates and the next cogging is processed. Cutting tools installed on the tool turret includes outer blade, middle-outer blade, middle-inner blade and inner blade. There are three main methods to check the machining accuracy of CNC machine tools [31]: sample detection method, indirect detection method [32], and direct detection method.

1.2 Energy consumption optimization

The energy-saving process optimization of CNC machine tools usually divides the machine tool's energy consumption into several parts [33]; the auxiliary system, the main drive system, the feed system, and the process of cutting and load. Gaussian process regression models are established according to these five parts, and the differential evolution algorithm is used to optimize the model. According to the processing requirements, a model of the energy consumption, cost, and time of the cutting process is established. The dynamic multi-swarm particle swarm optimization algorithm is used for calculation to obtain more diverse and convergent results [34]. The energy consumption of CNC milling machines can be divided into three parts, namely fixed energy consumption, no-load energy consumption, and milling energy consumption [35]. It can analyze the multi-source energy flow of the machine tool and the energy consumption of the machining process to establish the power of the machining stage equations and energy consumption estimation models. The calculation of energy consumption can be carried out from the amount of cutting, processed gears, cutting tools, cutting fluids of CNC machine tools, etc. to study carbon emissions [36]. The finished processing time is then established. The machining surface accuracy and other conditions are based on the reduction of carbon emissions from the spindle speed and feed rate during machining. A multi-objective optimization model for optimizing carbon emissions and processing costs can be established [37], which can be solved by genetic algorithms [38], and the feasibility of the optimization method of the model is verified by simulation calculation. On the digital intelligent machine tool, a system for detecting the energy consumption of the machining process of the CNC machine can be designed based on the upper computer information interaction unit and the lower computer information acquisition element.

1.3 Process optimization

The process design optimization of CNC machine tools has a significant impact on their precision performance. The design of the machining process of the machine tool should determine the processing steps and clamping methods of the workpiece by analyzing the mechanical drawings of the parts; the geometric elements of the outline of the part, the accuracy requirements of the dimensional tolerances, the accuracy requirements of the shape and position tolerances, surface smoothness requirements, material quality requirements, and the number of processes mode [33]. The feed route, choice of cutting amount, and choice of tool can then be determined. In order to prevent the tool from colliding with the workpiece during the machining process, the optimization of the machining process must be carried out by interference avoidance research [39, 40]. Based on the algorithm of coordinate extremes, the complex surface should be simplified by taking an arc of the surface [41], and performing measurement on the bisector. The vibration of the machine tool gravely affects its machining accuracy [42]. The hardness and thickness of the workpiece during the machining process [43] and the force of the machine tool are taken into consideration [44]. For the movement path [45], the movement speed [46] is optimized to achieve the machine vibration damping control. Another vibration reduction method is passive vibration reduction, which using materials with strong vibration resistance strengthens the stability of CNC machine tools [47]. Another factor affecting the machining accuracy of CNC machine tools is thermal deformation. During the working process of the machine tool, the moving parts of the machine tool will be affected by thermal deformation, which will cause relative displacement between the tool and the workpiece [48]. The solution is to strengthen cooling and lubrication during the work process to reduce the displacement [49]. The thermal characteristics, machining environment, and specific cutting parameters of the machine tool determine the size of the thermal error by solving the function of the time-varying temperature field under given conditions [50]. The purpose of the auxiliary heat source is to balance the temperature field to reduce heat source interference. In short, the errors caused by structural deformation, vibration, and high temperature in CNC machine tools can be compensated with the following methods: single error synthesis compensation technology [51], geometric error direct compensation technology, geometric error synthesis compensation technology [14], single-term error synthesis compensation technology by studying the error produced by a certain CNC machine tool, and geometric error direct compensation technology by measuring the error data to directly error compensation on the CNC machine tools. Using geometry error synthesis compensation technology, the unidirectional error information is decomposed by obtaining the synthetic error value of the CNC machine tool. It is particularly important to emphasize the tremendous difficulty in accurately measuring the angular error of the spindle during the general process; however, it can now be solved by a matrix using a laser interferometer [52]. For complex curved surface processing, multi-step compensation must be adopted as follows: (1) pre-compensation; (2) error detection; and (3) reverse compensation. For the deformation of the tool during machining, mirror image anti-deformation compensation that mirrors the tool position point and the tool axis vector must be adopted [53].

To sum up, the reliability of CNC machine tools is guaranteed by the processing quality of each part of the CNC machine tools, and the quality of processed parts is controlled by the quality of the processing procedures. For the analysis of the machining process of the machine tool, it is necessary to analyze the machining process of a key part. The optimization parameters of CNC machine tool machining process are mainly energy consumption during machining, machining efficiency, and machining accuracy. Due to the limited space, this chapter only discusses the influence of optimization parameters on machining accuracy.

2. CNC common processing optimization approach

The Computer Numerical Control (CNC) machine tools lie in the working principle of CNC machine tools: according to the requirements of part processing and using CNC language, the processing sequence is written into parameter programs, and the program is analyzed by Multipoint Control Unit (MCU). After processing, the execution instructions are sent to the Driving System of step motors or Servo motors and the machine tool will start machining the work piece. The position and velocity signals of the machine tool will then be sent back to the MCU. The schematic of CNC is shown in **Figure 2**.

From a mechanical point of view, CNC machine includes three main sub-systems: the main spindle sub-system, followed by the supporting sub-system and the conveying mechanism sub-system. From a control point of view, it mainly includes two sub-systems: Data Processing Unit and Control Loop Unit. The processing of parts action is completed through the numerical control system by controlling the mechanical structure of the CNC machine tool. The coordination between their functions is an important index for evaluating the machine tools under the CNC system. To fully utilize the merits of CNC for precision machining, a reasonable milling process should be selected according to different materials. Vibration caused by excessive rotation speed of the machine tool spindle will affect machining accuracy. The outer ring can be processed after specific processing. During the process of boring bar technology, thermal stiffness enhancement technology can be used to improve the rigidity and strength of the material.

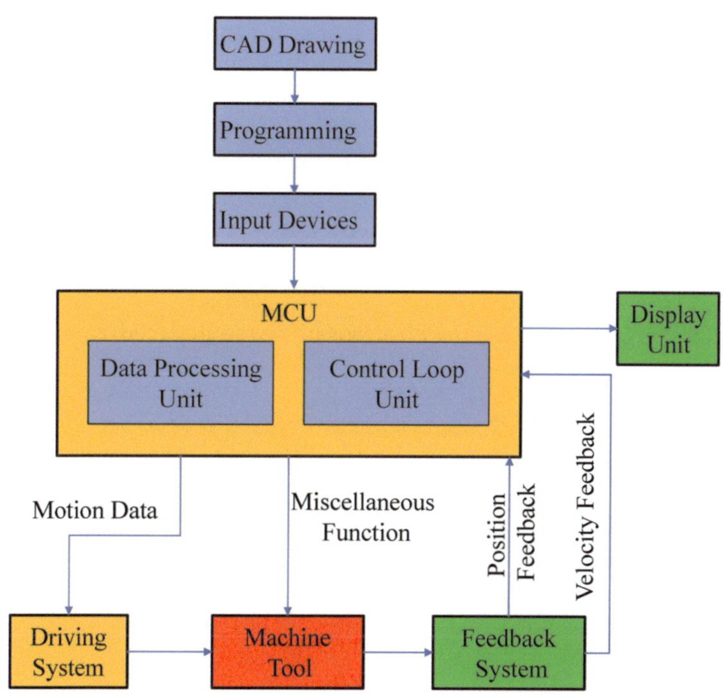

Figure 2.
Schematic diagram of CNC.

2.1 Cutting force monitoring

To obtain the wear and tear errors in the manufacturing and processing of cutting tools, we need to monitor them in real time to detect problems in time, check, integrate, and summarize them, and then establish them through the data system error compensation model. The deformation of the machine tool due to insufficient static stiffness, which causes deviations between the actual position and the ideal position of the tool and the workpiece, seriously affects its machining accuracy. In Precision Machining process on CNC, the cutting force in tangential and radial directions needs to be predicted based on monitoring signals and machining process parameters such as cutting depth, cutting width, rotational speed of the main spindle, and feed speed of the cutting tool. The experimental data of the cutting forces are shown in **Figure 3**.

2.2 Processing path optimization

The kinematic configuration of CNC machine tools directly affects the nonlinear errors generated in the process of free-form surface machining, and has a direct impact on processing energy consumption and design of the supporting CNC system.

After determining the motion configuration of the CNC machine tool, the mechanism of each feed axis has a significant impact on the overall rigidity, positioning error, maintainability, and other indicators of the CNC machine tool, which determine its machining accuracy and work reliability. For instance, in order to machine complex surface S12, it needs to be decoupled into surface S1 and S2. According to the required feed motion of surface S1 and S2, we can get the feed motion of combined curved surface S12 as shown in **Figure 4**.

A heuristic algorithm (HA) is one that is designed to solve a problem in a faster and more efficient fashion than traditional methods at the expense of optimality, accuracy, precision, or completeness. Although it has the capability of convergence and obtaining the optimization result efficiently, it depends heavily on initial result and may only obtain local optimal solution of non-smooth curve functions. Moreover, it occupies a large amount of computing resources and has a poor real-time performance, which may be incompatible with Computer Aided Manufacturing (CAM) on CNC. In order to overcome this issue, a discrete tool planning is proposed as follows:

$$Path = \left\{(A,C)^1_{path_1}, (A,C)^2_{path_2}, (A,C)^3_{path_3}, \dots (A,C)^m_{path_m}\right\}. \quad (1)$$

where, $(A,C)^m_{path_m}$ represents the positions of two rotary feed axes A and C when the cutting point is on m.

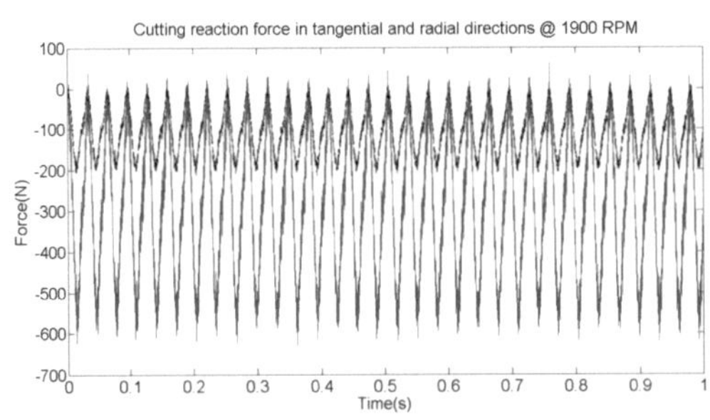

Figure 3.
The experimental cutting forces in tangential and radial directions.

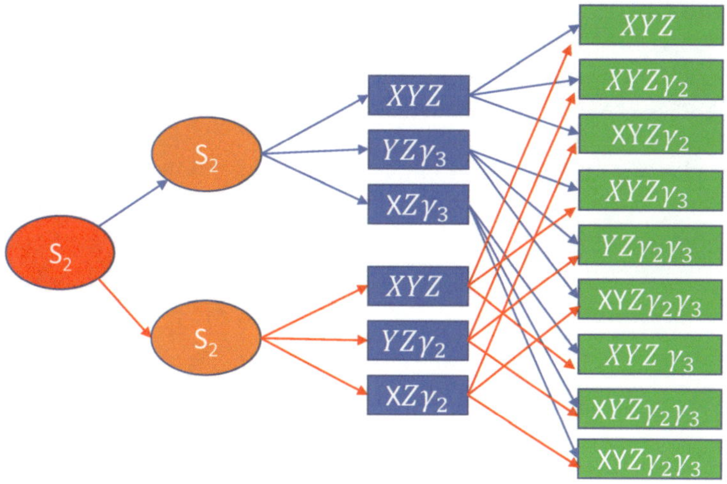

Figure 4.
The feed motion of S12 obtained from that of decoupled surface S1 and S2.

One methodology for positioning error compensation of CNC machine tools is through interpolation. It produces third-order splines from the coordinates of discrete points measured by Laser Doppler velocimetry (LDV). Between $[x_i-1, x_i]$, the interpolation function of splines $s(x)$ can be deduced as

$$s(x) = \frac{\left[(x_i - x)^3 * M_{i-1} + (x - x_{i-1})^3 * M_i\right]}{6h_i} + \left(y_{i-1} - \frac{h_i^2}{6} * M_{i-1}\right)\frac{x_i - x}{h_i} + \left(y_i - \frac{h_i^2}{6} * M_i\right)\frac{x - x_{i-1}}{h_i}, \text{i}= 1, 2, \dots \text{n} \tag{2}$$

In order to obtain M_{i-1}, let

$$\begin{cases} d_i = \dfrac{6}{h_i + h_{i+1}}\left(\dfrac{y_{i+1} + y_i}{h_{i+1}} - \dfrac{y_i + y_{i-1}}{h_i}\right) = 6f(x_{i-1}, x_i, x_{i+1}) \\ u_i = \dfrac{h_i}{h_i + h_{i+1}}, \lambda_i = \dfrac{h_{i+1}}{h_i + h_{i+1}} = 1 - u_i \end{cases} \tag{3}$$

Since the binomial multi-order derivative is continuous, M_i satisfies $N - 1$ equations,

$$u_i M_{i-1} + 2M_i + \lambda_i M_{i+1} = d_i \; i = 1, 2, 3 \dots, n - 1 \tag{4}$$

The first type of boundary conditions is

$$S'(X_0) = f_0' = m_0, S'(X_n) = f_n' = m_n \tag{5}$$

Then, Eq. (4) can be present as

$$\begin{bmatrix} 2 & \lambda_0 & & & \\ \mu_1 & 2 & \lambda_1 & & \\ & \dots & \dots & \dots & \\ & & \mu_{n-1} & 2 & \lambda_{n-1} \\ & & & \mu_n & 2 \end{bmatrix} \begin{bmatrix} M_0 \\ M_1 \\ \dots \\ M_{n-1} \\ M_n \end{bmatrix} = \begin{bmatrix} d_0 \\ d_1 \\ \dots \\ d_{n-1} \\ d_n \end{bmatrix} \tag{6}$$

2.3 Processing dynamic profile optimization

To prevent CNC machining in normal condition, the acceleration and velocity of the moving stages have to be monitored and should be less than the maximum design range of CNC machine.

With the known coordinates of two adjacent track points (x_{n+1}, z_{n+1}), (x_n, z_n) and the velocity of the main spindle, the acceleration and velocity of point n are approximately.

$$\Delta t = \frac{\Delta\theta}{\omega}, v_n = \begin{bmatrix} \dfrac{x_{n+1} - x_n}{\Delta t} \\ \dfrac{z_{n+1} - z_n}{\Delta t} \end{bmatrix}, a_n = \begin{bmatrix} \dfrac{\dot{x}_{n+1} - \dot{x}_n}{\Delta t} \\ \dfrac{\dot{z}_{n+1} - \dot{z}_n}{\Delta t} \end{bmatrix}. \tag{7}$$

Figure 5 shows the monitored acceleration in x and z directions respectively. Their values should be within CNC machine design range. NC machining process usually faces "over cutting" issue, which can be minimized through the commonly used method of rotating the tool angle as shown in **Figure 6**.

The extreme value $S_t$and root mean square value S_q of the machining surface shape are the two major indicators of the machining accuracy and are calculated by the following equations:

$$S_t = \max(e) - \min(e), S_q = \sqrt{\sum\nolimits_{i=1}^{n} e_i^2/n}. \tag{8}$$

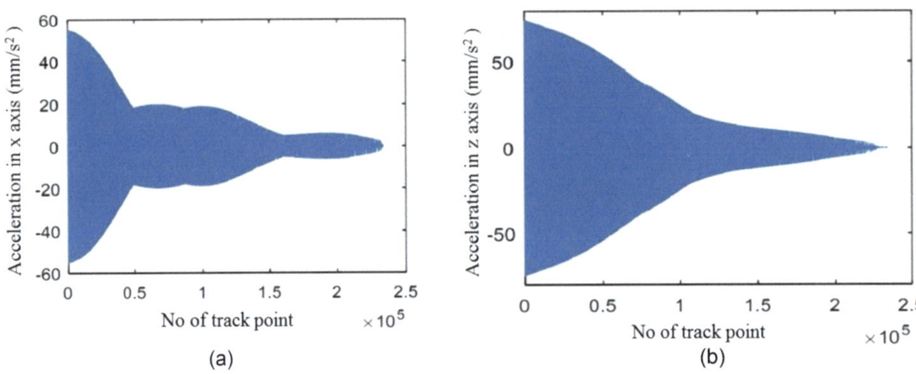

Figure 5.
Discrete point acceleration of machining trajectory. (a) Acceleration of track pint in x axis and (b) acceleration of track pint in x axis.

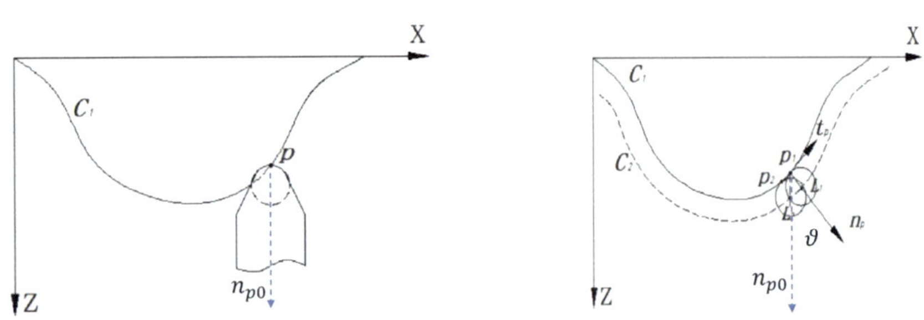

Figure 6.
Machining tool overcutting compensation.

where e represents the shape tolerance of the machining surface, N represents the number of the measured point.

Figure 7 shows the implementation of cutting tool compensation to reduce the machining error. It can be seen that tool compensation can effectively solve machining errors from 0.8 to 0.5 μm due to overcutting of tools through rotating cutting tool angle ϑ.

To minimize the value of S_(t)and S_(q), the optimized machining profile should be conducted as shown in **Figure 8**. In order to make the movement of the machine tool smooth and the processing more stable, the concept of "jerk" is introduced and made as constant. Then, the profile of acceleration is trapezoid. It can be seen that the speed profile of trapezoid acceleration is smoother than that of constant acceleration. As we know that the reaction force of the worktable and feed screw becomes infinite if the acceleration is suddenly changed. As a result, it may damage some motion transmission parts of CNC, such as worktable, lead screw, or servo motor.

Usually, in order to obtain high quality of the machining parts and extend the working life of the CNC machining, the speed of CNC is kept as constant as possible besides start and end stage of the machining processing.

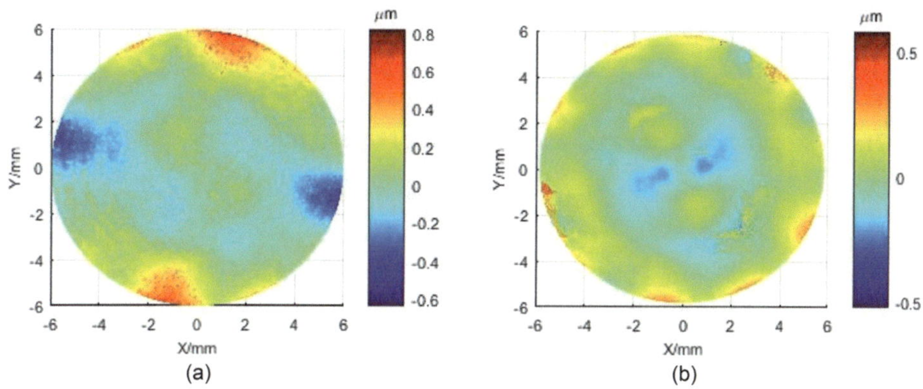

Figure 7.
Cutting tool compensation to reduce machining error. (a) Machining error before tool compensation and (b) machining error after tool compensation.

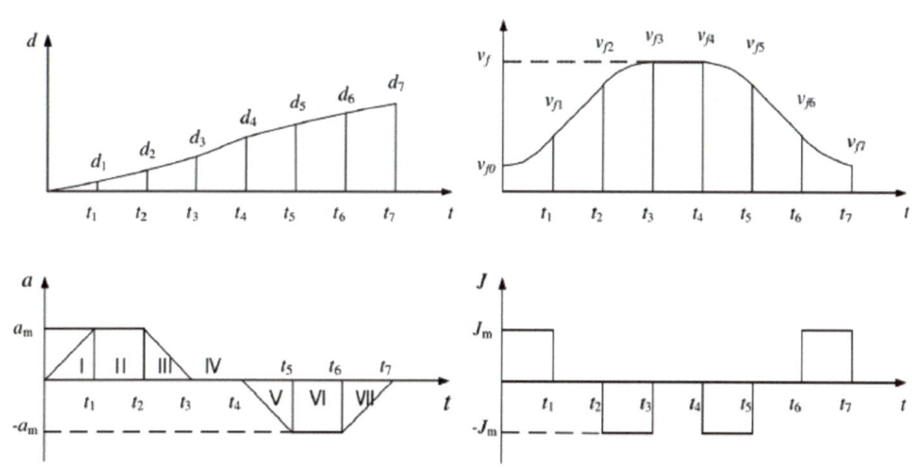

Figure 8.
Distance, velocity, acceleration, jerk vs. time profile in acc/dec process of CNC machining.

3. Processing tools deflection compensation

Any machining process will induce dynamic reaction forces on the cutting tool. Cutting tool deflection due to improper clamping, tool wear, and error in machine accuracy may lead to error in the final product. Hence, the current trend in the industry is to detect such errors and compensate for them in machining to avoid scraping the material afterward. In this chapter, we attempt to detect the cutting tool deflection while machining the workpiece and hence predict error on the workpiece. To detect the deflection during machining, we use either a contact or a non-contact sensor. One of the major considerations in choosing the right sensor is cost and ability to measure the error under harsh environments. In this section, compared to capacitive probe, we finally decide to adopt eddy current gap sensors to detect deflection of the cutting tool during machining of a simple profile and predict error on the workpiece. The predicted error was then verified using Coordinate Measuring Machine.

3.1 Displacement sensor for tool compensation

To compensate for the cutting tools or work piece deflection, there are two types of commonly used displacement sensors to monitor: (1) capacitance gap sensors and (2) Eddy current type gap sensors.

3.1.1 Capacitance gap sensors

The capacitance of a parallel-plate capacitor is given by

$$C = \frac{\varepsilon_r \varepsilon_0 A}{d} \tag{9}$$

where, ε_r represents electric constant of the insulating medium ($\varepsilon_r = 1$ for air); ε_0 represents permittivity of air or vacuum, which is 8.85×10^{-12} F/m; A represents overlapping area in plates; and d represents varying distance between the bottom surface of the electrode of capacitive probe and the surface of measured object (tool cutting tip).

The working schematic of the capacitive probe is shown in **Figure 9(a)**. It is evident that changes in the distance between the capacitive probe and the measured object (cutting tools) change the capacitance, which in turn changes the current flow in the sensing element. RF oscillator generates 500 kHz and 1 MHz high-frequency electric field to focus the sensing field on the cutting tool. **Figure 9(b)** shows the relationship between distance d and capacitance c. It shows d decreases as c increases. Based on this principle, the distance between plates can be determined by calculating the capacitance value.

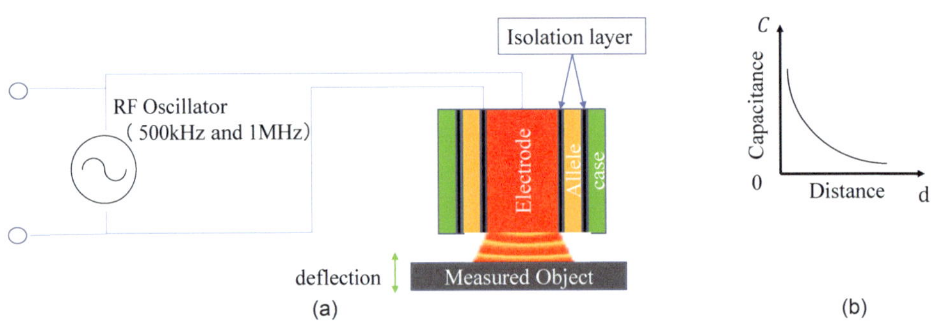

Figure 9.
The working schematic and principles of capacitive probe. (a) Schematic of capacitive probe and (b) capacitance vs. distance.

3.1.2 Eddy current type gap sensors

The working schematic of eddy current gap sensor is shown in **Figure 10**. When a nonmagnetic conductive target material is introduced into the coil field, eddy currents are induced on the target's surface. These currents generate a secondary magnetic field, inducing a secondary voltage in the sensor coil. The result is a decrease in the coil's inductive reactance (the coil-target interaction is similar to the field interaction between the windings of a transformer). By calculating the "effective impedance" of the sensor coil, the distance to the target can be determined.

Nonmagnetic conductive target materials (e.g., aluminum, copper, brass, gold), with low resistivity and a magnetic permeability of 1, can provide output sensitivity in terms of impedance change per unit of target displacement.

The capacitive probe is suitable for measuring any metal surface. Its measuring accuracy can reach up to several nanometers. However, the dusty working environment is not suitable for applying capacitive probe to measure the target, as the gap between probe and the measured target is contaminated by dust, liquids such as coolant. As a result, the measuring accuracy of the capacitive probe is affected. On the other hand, the current gap sensor is suitable for measuring conductive material except steel. However, it cannot measure thin materials because of insufficient eddy current generated, thus affecting the measuring accuracy of the gap sensor.

3.2 Tool deflection compensation approach

Figure 11 shows 5G AI Edge computing configuration for CNC cutting tool deflection compensation. The original signal is collected by sensor and sent to 5G AI computing for preprocessing such as noise reduction with low-pass filter, compressed data etc. The dimension-reduced data is sent for cloud computing through a 5G network router. The cloud computing device uses genetic programming (refer to **Figure 12**) to generate the cutting tool deflection compensation algorithm and sends it back to the 5G AI edge computing local network. 5G AI edge computing device then calculates the compensation value and sends it to MCU to compensate for the cutting tool deflection.

3.3 Algorithm adopted in tool deflection compensation

Genetic programming (GP) is adopted in this study to conduct tool deflection compensation. It evolves from computer programs that are traditionally represented in memory as tree structures. Trees can be easily evaluated in a recursive manner. Every tree node has an operator function and every terminal node has an operand, rendering mathematical expressions easy to evolve and evaluate. Thus, traditionally GP favors the use of programming languages that naturally embody tree structures.

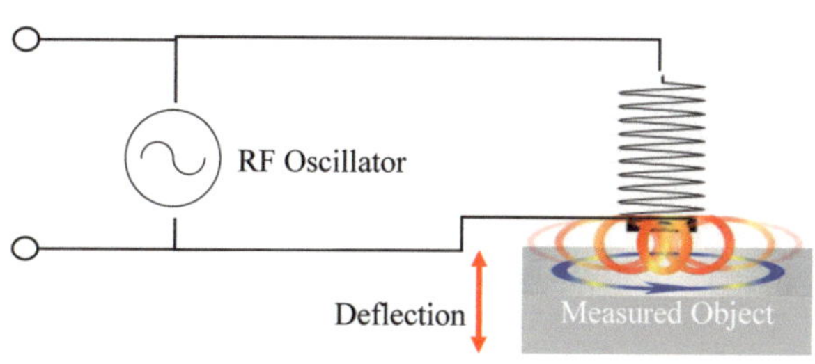

Figure 10.
The working schematic of eddy current gap sensor.

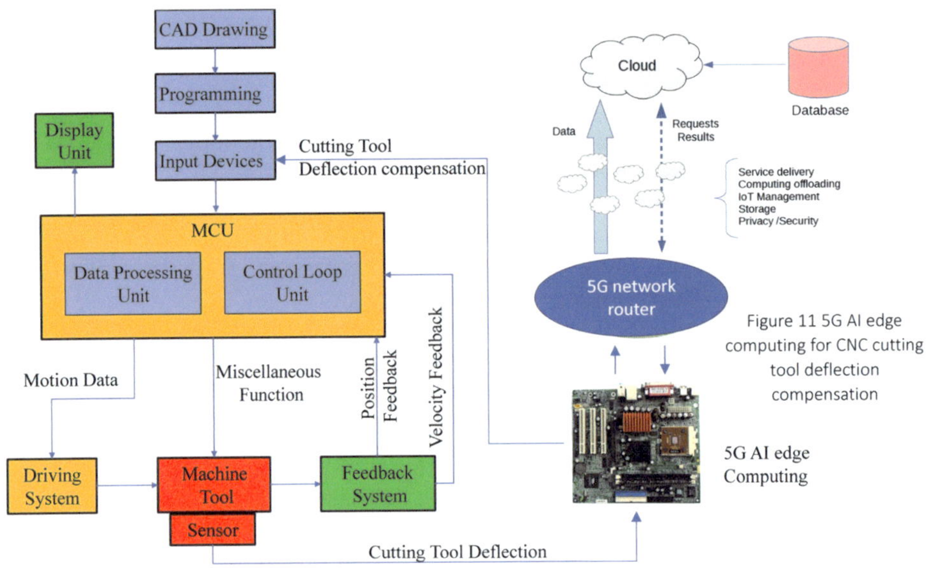

Figure 11.
5G AI edge computing for CNC cutting tool deflection compensation.

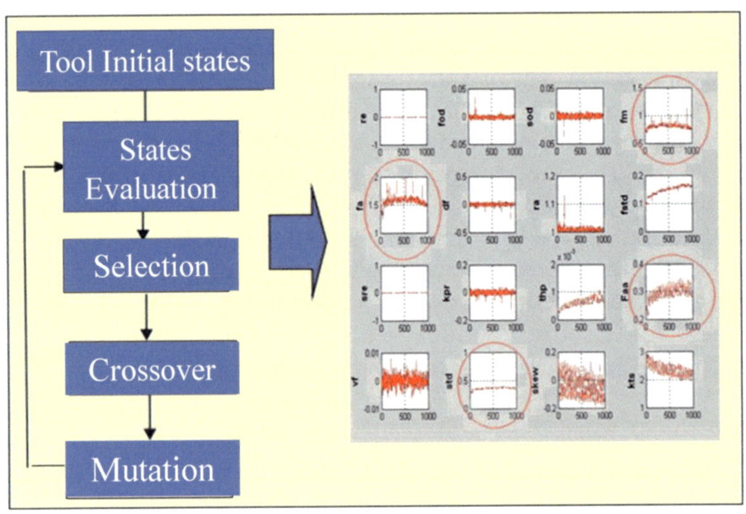

Figure 12.
Schematic of genetic programming for tool deflection compensation.

GPLAB is a genetic programming toolbox for MATLAB® initially developed by Sara Silva. The main modules of GBLAB are GENPOP, GENERATION, and SETVARS. GENPOP module generates the initial population and calculates its fitness based on default or user defined fitness function. Terminal set:

Terminal set: "X1," "rand," and "100".

Terminal condition: Maximum generation reached.

Function set: plus (+), minus (−), times (*) and other protected function in.

Fitness function: Fitness function is defined as the minimum value between actual value and predicted value. During the programming calculation process, the new population should be superior to the previous population. In other words, the new fitness value will be lower than previous one.

$$\text{Fitness} = \min\ (\text{abs}\ (\text{predicted result} - \text{expected result})). \tag{10}$$

The schematic of genetic programming for tool deflection compensation is shown in **Figure 12**. The optimal tool deflection compensation algorithm is obtained through hundreds of generation selection, crossover, and mutation operation until set fitness function is reached.

3.4 Tool deflection compensation experiment

The material of cutting tools includes high-speed steel, ELMAX chromium-molybdenum-vanadium alloy steel, or nickel-chromium alloy steel. However, the cutting chip will reduce the capacitive probe measurement sensitivity and accuracy and damage the capacitive probe. Therefore, the eddy current gap sensor will be adopted in this working environment. The deflection signal is sent to 5G AI edge computing for signal pre-processing, before sending for computing through genetic programming. The computing results are sent back to MCU of the CNC to control the movement of the main spindle and the worktable.

The cutting surface error compensation with and without cutting tool deflection compensation are shown in **Figure 13**. It shows that the maximum machining error

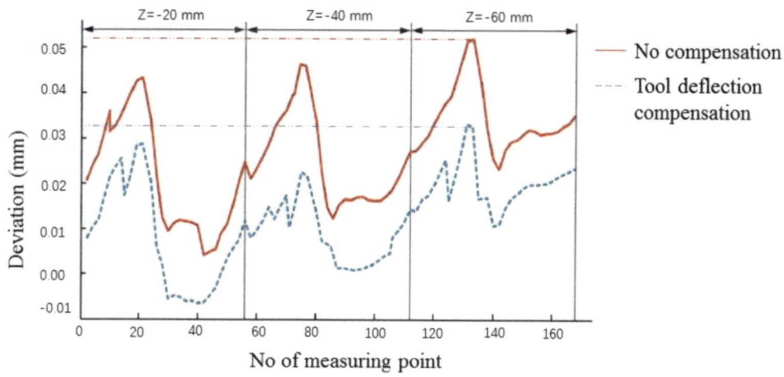

Figure 13.
Cutting surface error with/without tool deflection compensation.

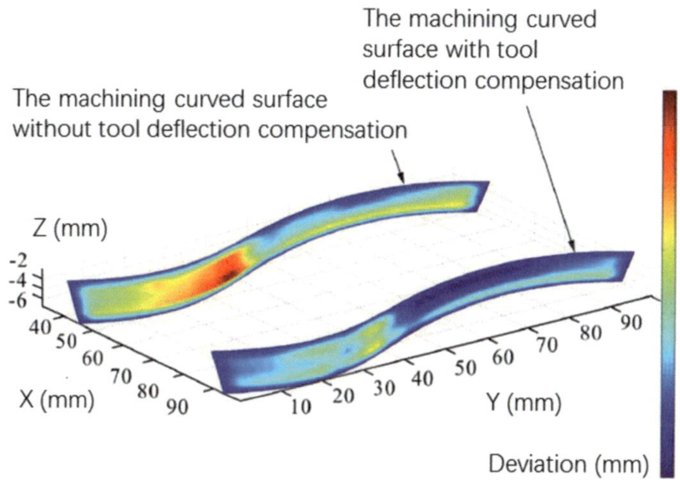

Figure 14.
The deviation distribution of the machining with/without tool deflection compensation.

is reduced from 0.053 to 0.032 mm, which is 40%. The deviation distribution of machining with/without tool deflection compensation is shown in **Figure 14** in three dimensions. It can be seen that at almost all measured points, the machine error on a curved surface without tool deflection is larger than that on a machining curved surface with tool deflection compensation. Moreover, the error is largest at the bottom area of inflection of the curved surface, reaching 0.053 mm. This may be due to two reasons. First, the cutting tool suffered the largest reaction force and induced the largest deflection. Second, the stiffness of the machining tool is smallest at the tip.

4. Processing error feedback compensation

4.1 Experimental setup of error feedback compensation

In order to further reduce the machining error after cutting tool compensation, the error feedback compensation strategy is selected as shown in **Figure 15**. Two accelerometers are attached to the side of the stator of the main spindle and the side of the workpiece to measure their vibration signals respectively. The vibration of the accelerator attached to the main spindle in idle and working conditions is monitored and feedback to CNC, then, the CNC control the main spindle rotation speed to obtain minimum vibration signal. The vibration sensor attached to the workpiece monitored the dynamical performance of the workpiece and the signal is sent back to the control unit for processing. A servo actuator is attached to the bottom surface of the workpiece to compensate for the machining error, which is based on control signal received from control unit. An acoustic emission sensor is attached near the workpiece to monitor the acoustic noise and the fault processing condition. The original signal is sent to control main server for processing and pattern and status recognition. Lastly, the control command is sent from the main server to CNC machine to minimize the machine error.

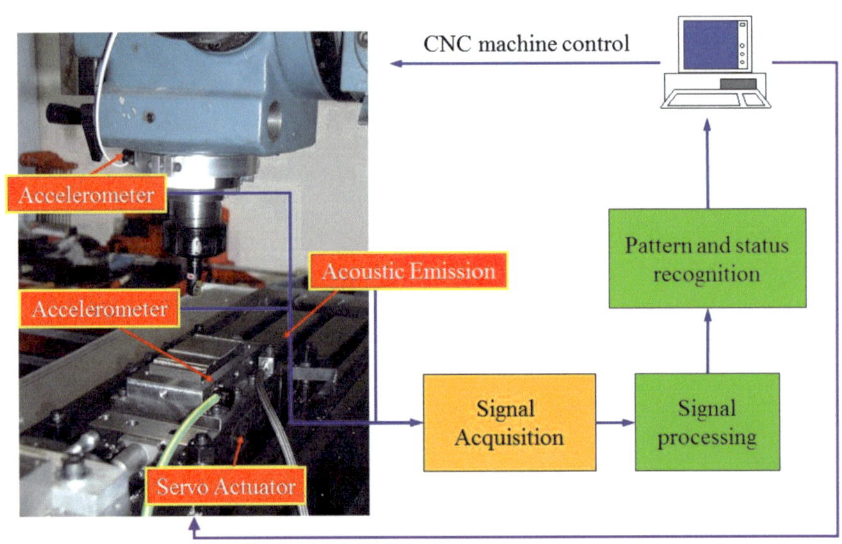

Figure 15.
Schematic of error feedback compensation.

4.2 Signal processing in error feedback compensation

Figure 16 shows error feedback compensation signal noise filter and reconstruction processing. The original time domain signal is obtained from the vibration sensor and converted into frequency domain through fast Fourier transform (FFT) algorithm. The bandpass filter is applied to filter out high frequencies and low-frequency and DC component. Interesting frequencies that include useful information are obtained after applying a signal technology to find peaks of the FFT in the range. The final step is to reconstruct the signal to control the actuator in order to compensate for machining error.

4.3 Experimental results of error feedback compensation

Figure 17 shows the cutting surface error without tool deflection compensation, with tool deflection compensation, and with error feedback compensation. It can be seen that the maximum machining error is reduced by 35% from 0.032 to 0.021 mm when the error feedback compensation is applied. Therefore, the total machining error is reduced by 60% from 0.53 to 0.021 mm. However, the error feedback strategy does not work at measurement point 40. Moreover, in two ranges (point 20–28, point 45–55), machining error is larger than that with tool deflection compensation strategy only.

The deviation distribution of machining without compensation, with tool deflection compensation, with error feedback compensation is shown in **Figure 18**. It can be seen that the deviation distribution of the machine curved surface with feedback compensation is minimal. On the other hand, the deviation distribution of the machine curved surface without tool deflection and error feedback compensation is largest. In other words, in order to achieve good machining quality product, the tool deflection compensation and error feedback compensation strategy should be adopted.

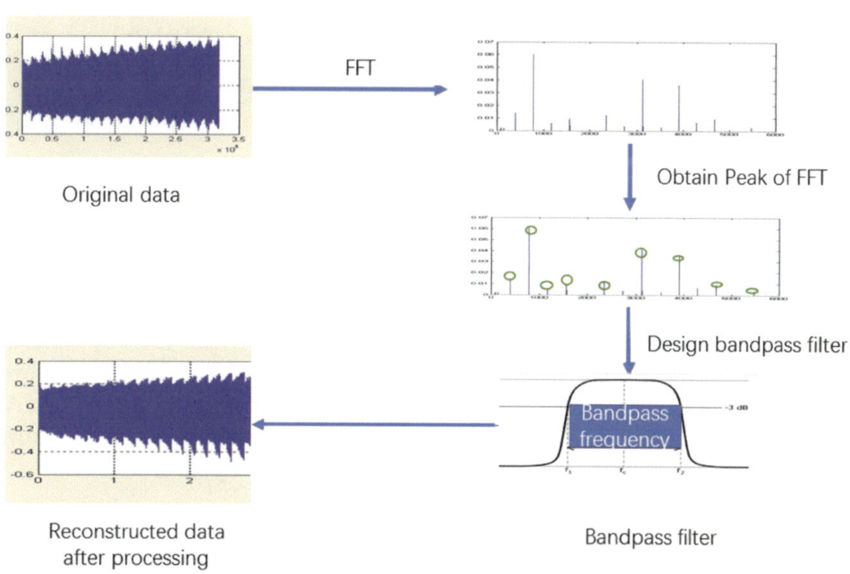

Figure 16.
Error feedback compensation signal noise filter and reconstruction processing.

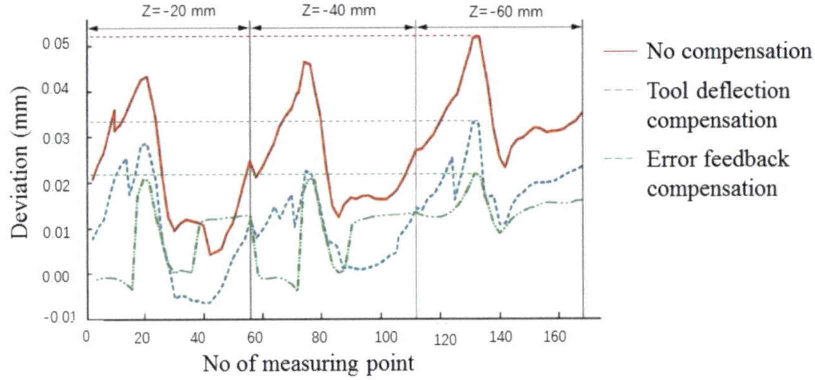

Figure 17.
The machining surface error without tool deflection compensation, tool deflection compensation, error feedback compensation.

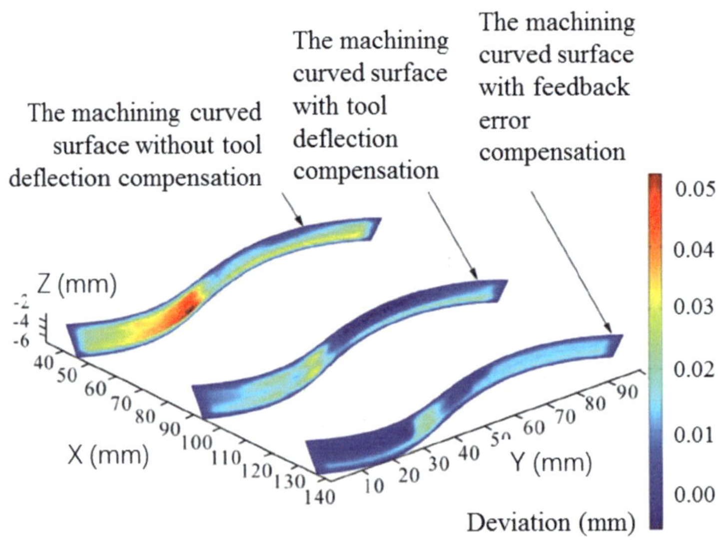

Figure 18.
The deviation distribution of the machining without compensation, with tool deflection compensation, with feedback error compensation.

5. Processing parameter optimization compensation

An experienced operator can significantly reduce machining error by applying cutting tool deflection and error feedback compensation when the optimized operation parameter is known. However, the process obtaining optimized operation parameter by try and error is time-consuming for new operator. Therefore, this section will introduce the parameter base optimization operation strategy. 5G AI edge computing and cloud computing strategy will be applied to obtain optimized operation parameter automatically.

5.1 Network configuration for processing parameter optimization

Figure 19 shows the cloud-based intelligent manufacturing configuration with 5G AI edge computing, which is also known as smart manufacturing configuration. The company data center managed by the IT department acts as a data transfer

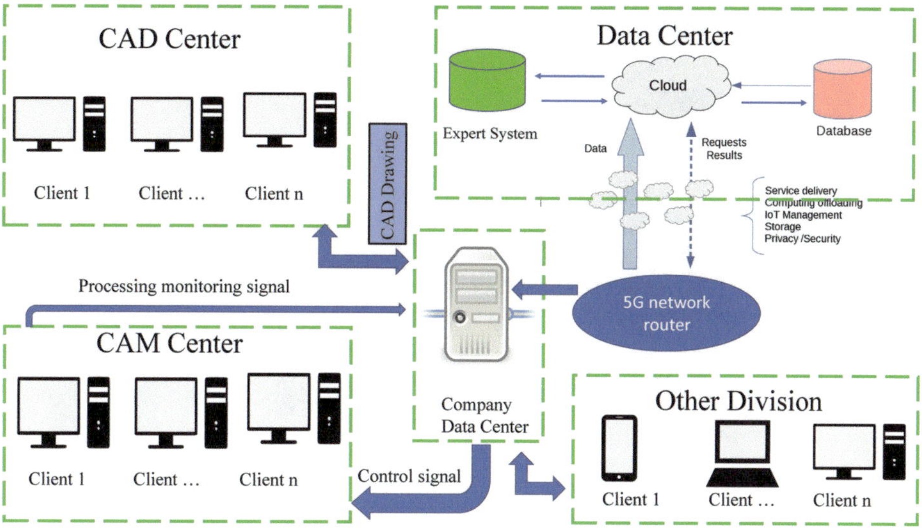

Figure 19.
The cloud-based intelligent manufacturing configuration.

bridge connecting each department of the company such as CAD center, CAM center, sales department, supply department, warehouse, maintenance department etc. It also acts as a data communication buffer linked to cloud computing centering out of the company. The designer can access the local data center to download the CAD drawing in order to modify the existing drawing, save the new CAD drawing, or update drawing to local data center. In the CAM center, every client computer is connected to a CNC machine. The CNC machine processing monitoring signal and the corresponding machine number information are sent to the company data center and to the cloud computing data center if necessary. All information needed for CAM center to produce the machining parts is sent through company data center as well. For instance, the production manager can download the CAD drawing from company data center and convert it to CAM file format. The technical person from the company then sends a file with control command or a file with the new compensation algorithm trained by cloud computing to the client PC in the CAM center. Cloud computing center receives request and the training data set from the company and sends back the proposed processing optimized parameters or processing control algorithm. In the cloud data center, the data received from the company will be stored in a database. It utilizes GP, MNNR, or other deep learning algorithm to obtain the optimized parameters through the expert system.

Figure 20 shows the cloud-based intelligent manufacturing configuration in the CAM center. The whole system configuration includes CNC real-time machining status monitoring, real-time environmental monitoring, CNC cutting tool deflection compensation, error feedback compensation as well as CNC machining operation process optimization through a combination of cloud computing and 5G AI edge computing. Unlike traditional standalone CNC machining, all CNC machines (refer to **Figure 21**) are connected through a local network client PC with cable or wireless in intelligent manufacturing environment. CAD drawing is sent to a local client instead of programming. The signals from sensors monitoring the CNC machining status are sent to 5G AI edge computing devices to pre-process. Processed data and corresponding machine no. are sent to the company data center through the client PC.

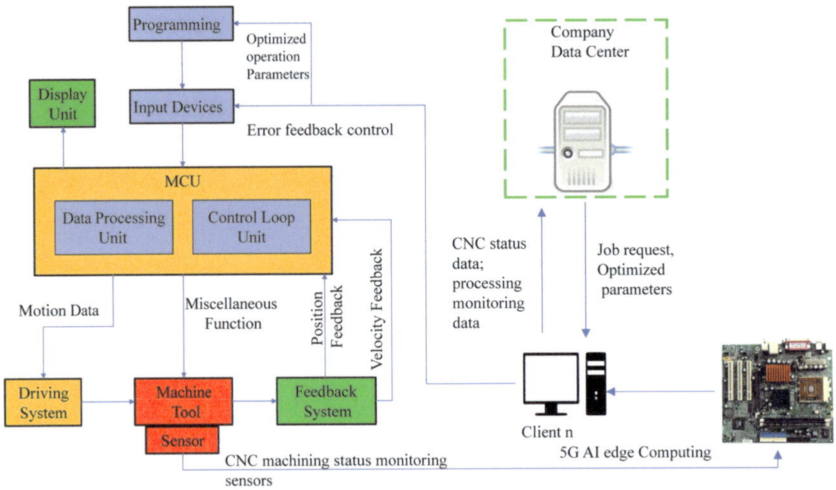

Figure 20.
The cloud-based intelligent manufacturing configuration in CAM center.

Figure 21.
A group of CNC machines connected through local networks.

5.2 Data collection for processing parameter optimization

Figure 22 shows a CNC machine connected to the local network. In order to obtain vibration signal to train the GP, Neural Networks (NNR), and Support Vector Machine (SVM) algorithm, two wireless accelerometers, Sensor 1 and Sensor 2, are attached to the cutting tool. Their signals will be sent to the cloud for GP, NNR, and SVM base computing. Specimen (Nylon) is clamped into three-jaw chuck. Vibration signal is sent to cloud server through 5G AI Edge computing device and 5G router. In order to find optimized operational rotational speed of main spindle, three groups of specimens (30 pcs) are machined with varied rotational speed as shown in **Tables 1–3**.

In order to get reliable data, the three groups of specimens must have the same experimental condition; the diameter and length of all experimental specimens are 25 and 30 mm respectively.

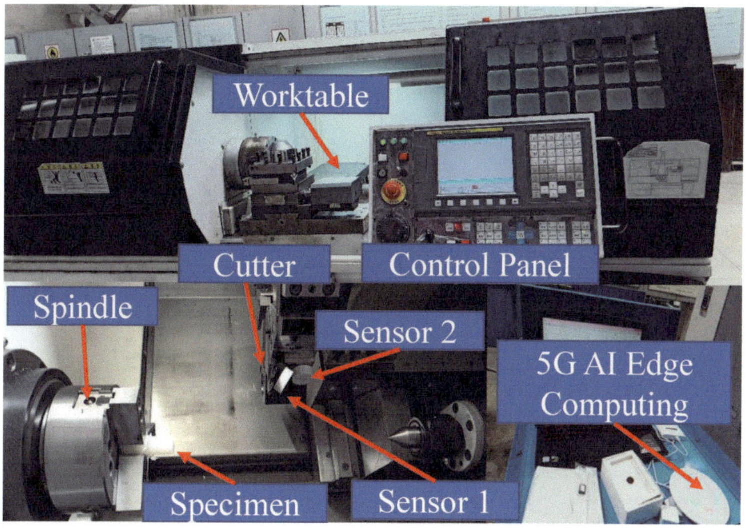

Figure 22.
Real-time monitoring of the CNC process in a production environment.

S/N	Diameter(mm)	Rotating speed (RPM)	Feed rate (r/min)	Length(mm)
A1	25	100	0.30	30
A2	25	300	0.20	30
A3	25	500	0.10	30
A4	25	700	0.08	30
A5	25	900	0.06	30
A6	25	1100	0.05	30
A7	25	1300	0.05	30
A8	25	1500	0.03	30
A9	25	1700	0.02	30
A10	25	1900	0.01	30

Table 1.
Group A specimens with varied machining parameters.

When the rotating speed of the main spindle of CNC machine is the same, the same feed rate is selected as well. The feed rate is increased with increasing rotating speed.

Figure 23 shows three groups of specimens labeled as A1–A10, B1–B10, and C1–C10 respectively. Group A specimens are used as training data set while group B and C specimens are used as testing data set.

5.3 Signal processing technologies for processing parameter optimization

There are various signal processing technologies for analyzing the monitoring machine working condition and product processing. The commonly used technologies are: time domain signal processing; FFT; wavelet technology; and short-time Fourier transform (STFT). Normally, time domain data are the original data recorded by the sensor, FFT is the technology developed to convert time domain

S/N	Diameter(mm)	Rotating speed (RPM)	Feed rate (r/min)	Length(mm)
B1	25	100	0.30	30
B2	25	300	0.20	30
B3	25	500	0.10	30
B4	25	700	0.08	30
B5	25	900	0.06	30
B6	25	1100	0.05	30
B7	25	1300	0.05	30
B8	25	1500	0.03	30
B9	25	1700	0.02	30
B10	25	1900	0.01	30

Table 2.
Group B specimens with varied machining parameters.

S/N	Diameter(mm)	Rotating speed (RPM)	Feed rate (r/min)	Length(mm)
C1	25	100	0.30	30
C2	25	300	0.20	30
C3	25	500	0.10	30
C4	25	700	0.08	30
C5	25	900	0.06	30
C6	25	1100	0.05	30
C7	25	1300	0.05	30
C8	25	1500	0.03	30
C9	25	1700	0.02	30
C10	25	1900	0.01	30

Table 3.
Experimental specimen group C with varied machining parameters.

data to frequency domain data. The wavelet technology is commonly used to decompose a complex signal to sleeve sub-level simple signals. STFT is the signal processing technology developed to overcome the drawbacks of FFT. In most cases, it is used to monitor the transition signal. In this section, each signal technology will be discussed based on experimental data. The purpose is to find out the most suitable signal processing technology to develop process parameter optimization.

Sensor 1 is attached to the tip of the cutter to monitor the cutting dynamic force-induced vibration. The sensor is a wireless accelerometer and powered by a built-in battery. As shown in **Figures 24** and **25**, time domain signal is repeatable with signal noise. However, we cannot trace enough useful information from the signal in time domain.

Due to lack of useful information from time domain data, frequency domain data are converted from time domain data by employing FFT technology. In machine fault diagnosis application, the fault frequencies can be found when the machine is under faulty working condition such as bearing wear and tear.

Figures 26 and **27** show the frequency domain data of Sensor 1 in X and Y directions, respectively. The domain frequency is the 1x order signal frequency, which is around 31.67 Hz. The second contribution signal is around 263 Hz. The third and fourth ones are that of 2x order and 3x order signal respectively.

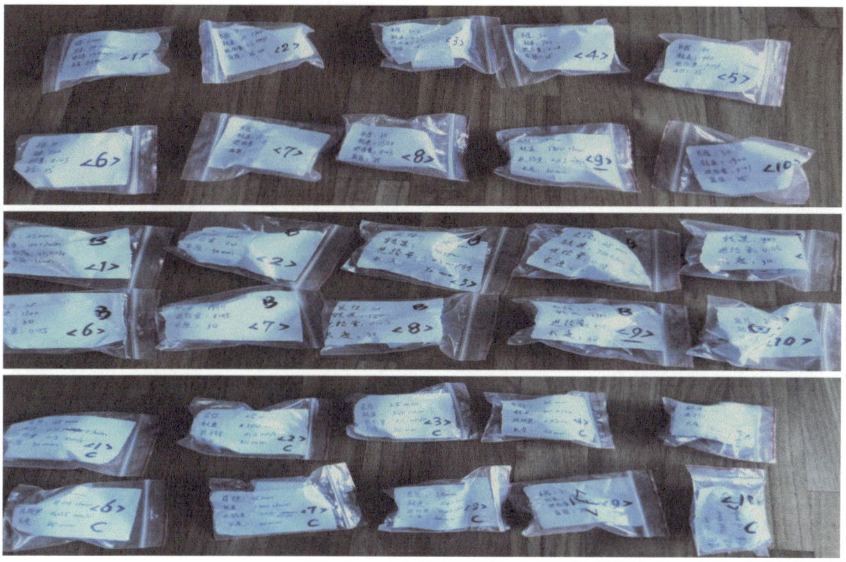

Figure 23.
Experimental specimen groups with varied machining parameters.

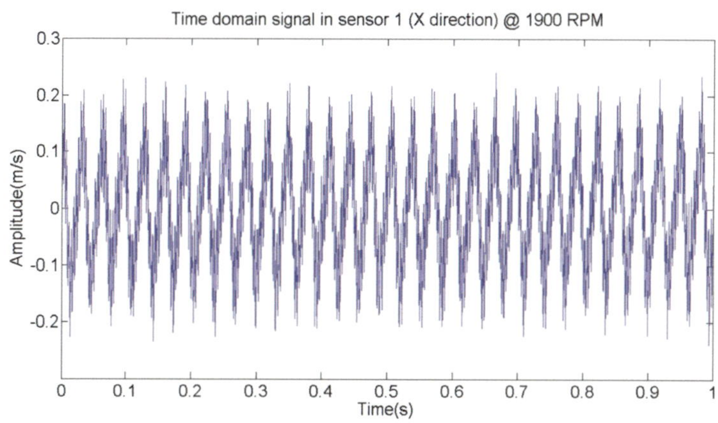

Figure 24.
Time domain signal in Sensor 1 in X direction.

Wavelet technology is originally developed to decompose complex signal. It may be used in the processing application to trace useful information. **Figure 28** shows wavelet spectrum in sensor 1 in X and Y directions; it can be seen that the signal is decomposed. However, only first two domain signals can be displayed. Moreover, the frequency of each signal cannot be recognized, which means this signal technology may be not suitable for processing optimization application. Other signal technologies should be further studied to be applied in this application field.

Figure 29 shows short-time Fourier transform spectrum in sensor 1 in X and Y directions. It can be seen that the dominant signal is the low-frequency component. However, it cannot provide any useful information for optimizing processing parameters. In **Figures 26** and **27**, we know that the dominant frequency is the 1x order of main spindle rotating frequency. Waterfall technology is used to apply FFT in several steps of the main spindle's rotational speed in order to trace any changes due to change in rotating frequency.

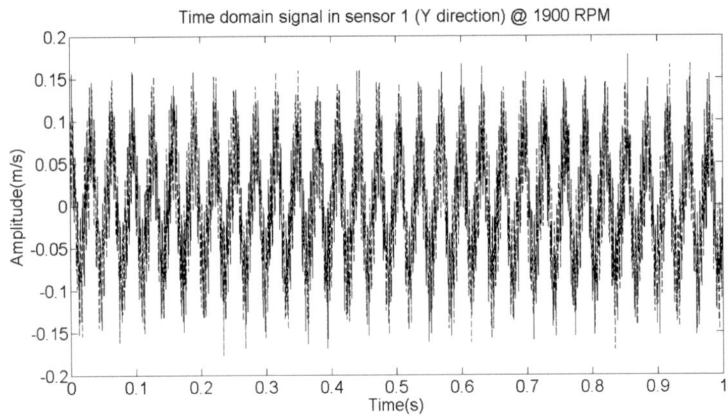

Figure 25.
Time domain signal in Sensor 1 in Y direction.

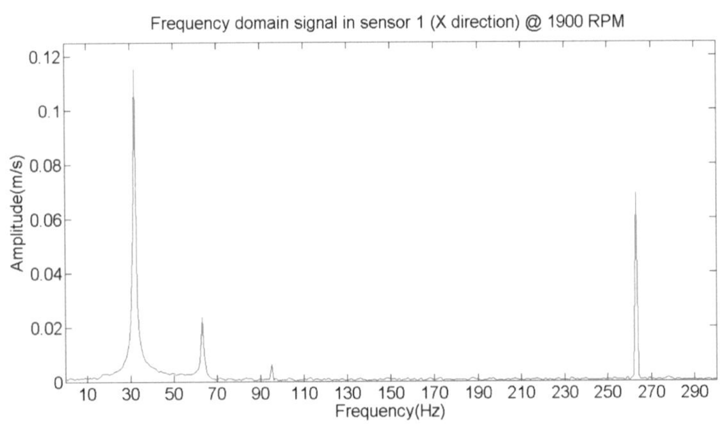

Figure 26.
Frequency domain signal in Sensor 1 in X direction.

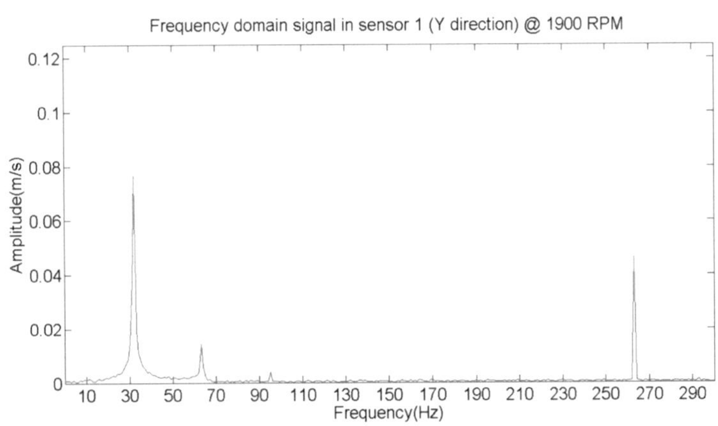

Figure 27.
Frequency domain signal in Sensor 1 in Y direction.

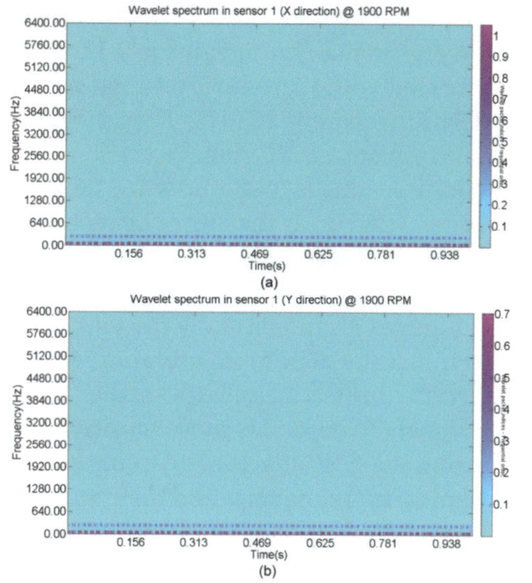

Figure 28.
Wavelet spectrum in Sensor 1 in X and Y directions.

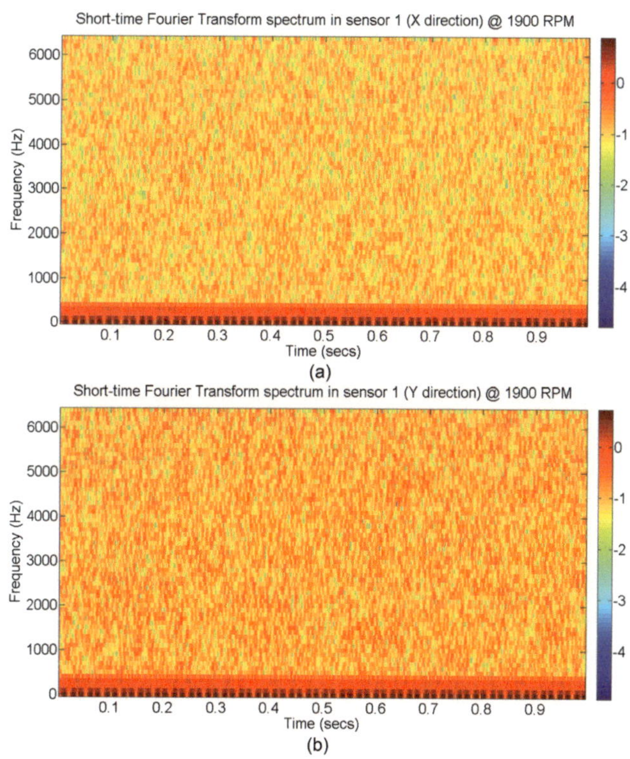

Figure 29.
Short-time Fourier transform spectrum in Sensor 1 in X and Y directions.

5.4 Experimental results of processing parameter optimization

The useful information is traced when the waterfall of speed spectrum is developed. Sensor 1 is attached to the tip of the cutting tool to monitor the cutting dynamic force-induced vibration. **Figures 30** and **31** show the vibration in X and Y directions respectively. In **Figure 30**, it can be seen that the dominant frequency components are 1x, 2x, and 3x. Moreover, the maximum value is 0.12 mm/s at 31.6 Hz and the vibration spectrum is increased with increasing rotating speed. The lowest resonance frequency of the CNC machine is 263 Hz. Therefore, the machine rotating speed has to deviate far away from 15,780 rpm.

In **Figure 31**, the maximum value is 0.08 mm/s at 31.6 Hz. Under 1900 rpm, the maximum value is smaller at 31.6 Hz, as the cutter bears bigger cutting force due to removal of material by cutter in X direction.

Sensor 2 is attached to the base of the cutting tool to monitor the cutting dynamic force-induced vibration. **Figures 32** and **33** show the vibration in x and y direction respectively. It can be seen that the maximum vibration of Sensor 2 is 0.025 and 0.16 mm/s in x and y direction respectively, both smaller than that of Sensor 1. Therefore, the vibration signal of Sensor 1 is more sensitive and is a better candidate than Sensor 2 for monitoring the CNC machine working status. Furthermore, only the frequency domain dominates frequencies and their amplitudes are selected to be sent to the cloud in order to reduce data size and increase data transmission rate.

After the speed spectrum is obtained, the frequencies and their peak value can be obtained by applying data mining methodology. **Figure 34** shows the tracing peak

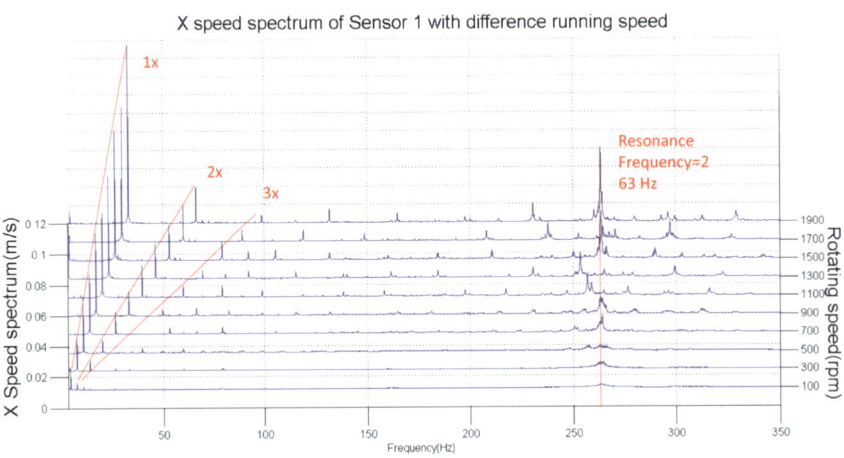

Figure 30.
Vibration signal in X direction of Sensor 1.

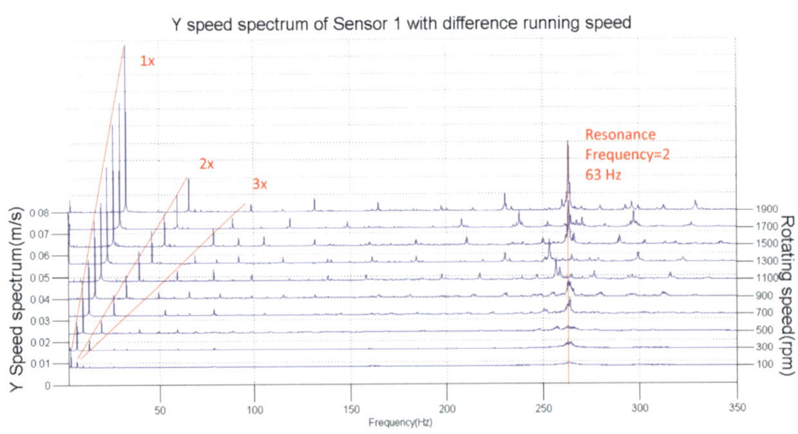

Figure 31.
Vibration signal in Y direction of Sensor 1.

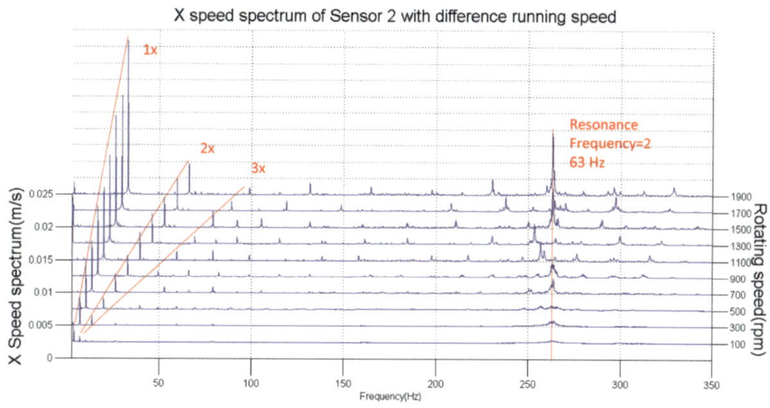

Figure 32.
Vibration signal in X direction of Sensor 2.

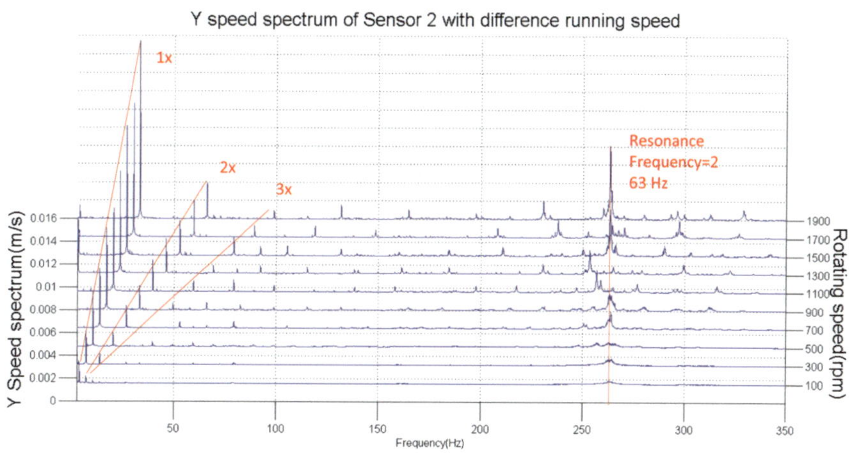

Figure 33.
Vibration signal in Y direction of Sensor 2.

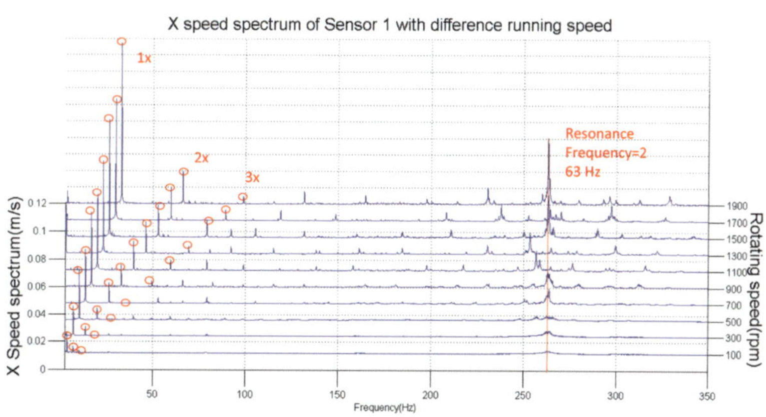

Figure 34.
Tracing peak values of group a specimens in frequency domain.

values of group A specimen in frequency domain with interesting 1x, 2x, and 3x data. The peak values of specimen A, B, and C are shown in **Tables 4–6** respectively.

The experimental parameters such as material, dimensions, temperature, reaction forces, vibration data, and machining quality such as tolerances will be sent to the Multilayer Artificial Neural Network (MANN) to train the optimized operation parameters. The MANN is shown in **Figure 35**.

Figure 36 shows results of MANN training based on training data, the training correction is about 99%. If more training data are adopted, the trained MANN should be more accurate.

Figure 37 shows general data classification results based on deep learning; there are four types of results: (a) The machining result is linearly classified into two

S/N	Rotating speed (RPM)	1st vibration in X (mm/s)	2nd vibration in X (mm/s)	3rd vibration in X (mm/s)	1st vibration in Y (mm/s)	2nd vibration in Y (mm/s)	3rd vibration in Y (mm/s)
A1	100	0.01349	0.00433	0.001583	0.00899	0.00288	0.001055
A2	300	0.02196	0.00683	0.001353	0.01464	0.00455	0.000189
A3	500	0.03635	0.00834	0.003509	0.02424	0.00566	0.0001864
A4	700	0.03911	0.01471	0.00283	0.02608	0.00981	0.000159
A5	900	0.05354	0.01576	0.004258	0.03569	0.01051	0.002839
A6	1100	0.05495	0.02097	0.006776	0.03663	0.01398	0.004519
A7	1300	0.06709	0.02301	0.005445	0.04473	0.01534	0.003630
A8	1500	0.08472	0.02256	0.01205	0.05661	0.01504	0.008036
A9	1700	0.08787	0.02442	0.007477	0.05858	0.01628	0.003930
A10	1900	0.11650	0.02332	0.005225	0.07771	0.01424	0.003484

Table 4.
Vibration speed spectrum of group A specimen.

S/N	Rotating speed (RPM)	1st vibration in X (mm/s)	2nd vibration in X (mm/s)	3rd vibration in X (mm/s)	1st vibration in Y (mm/s)	2nd vibration in Y (mm/s)	3rd vibration in Y (mm/s)
B1	100	0.01227	0.00393	0.001389	0.00856	0.00289	0.000955
B2	300	0.01996	0.00625	0.001356	0.01392	0.00457	0.000159
B3	500	0.03305	0.00768	0.003509	0.02375	0.00522	0.000176
B4	700	0.03556	0.01389	0.003871	0.02548	0.00919	0.000168
B5	900	0.04867	0.01435	0.006163	0.03446	0.00985	0.002762
B6	1100	0.04996	0.01906	0.004949	0.03438	0.01356	0.004327
B7	1300	0.06099	0.02091	0.005445	0.04293	0.01415	0.003403
B8	1500	0.07723	0.02051	0.005691	0.05308	0.01438	0.007534
B9	1700	0.07989	0.02221	0.005377	0.05492	0.01526	0.004673
B10	1900	0.10590	0.02128	0.004725	0.07284	0.01458	0.003266

Table 5.
Vibration speed spectrum of group B specimen.

S/N	Rotating speed (RPM)	1st vibration in X (mm/s)	2nd vibration in X (mm/s)	3rd vibration in X (mm/s)	1st vibration in Y (mm/s)	2nd vibration in Y (mm/s)	3rd vibration in Y (mm/s)
C1	100	0.013776	0.004561	0.001615	0.00969	0.00309	0.001135
C2	300	0.022455	0.008512	0.001489	0.01597	0.00483	0.000160
C3	500	0.037152	0.009501	0.003332	0.02594	0.00595	0.000166
C4	700	0.039975	0.016095	0.003891	0.02748	0.01059	0.000168
C5	900	0.054656	0.021396	0.005345	0.03825	0.01259	0.003065
C6	1100	0.056174	0.020348	0.006918	0.03945	0.01489	0.004847
C7	1300	0.068568	0.021915	0.005555	0.04795	0.01645	0.003989
C8	1500	0.086655	0.020513	0.007123	0.06069	0.01668	0.008615
C9	1700	0.089672	0.024926	0.007635	0.06278	0.01756	0.005341
C10	1900	0.118923	0.023856	0.005332	0.08325	0.01569	0.003732

Table 6.
Vibration speed spectrum of group C specimen.

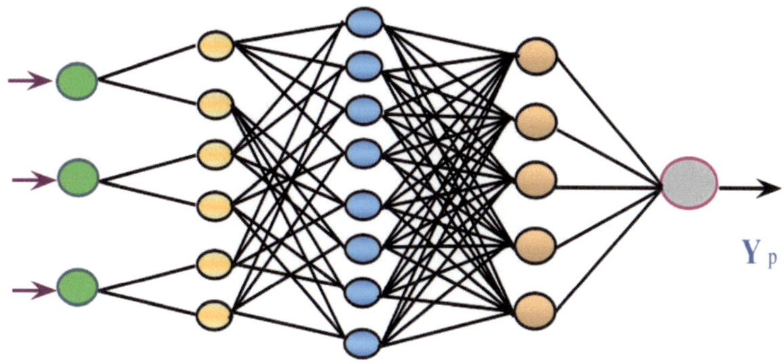

Figure 35.
Multilayer artificial neural network (MANN).

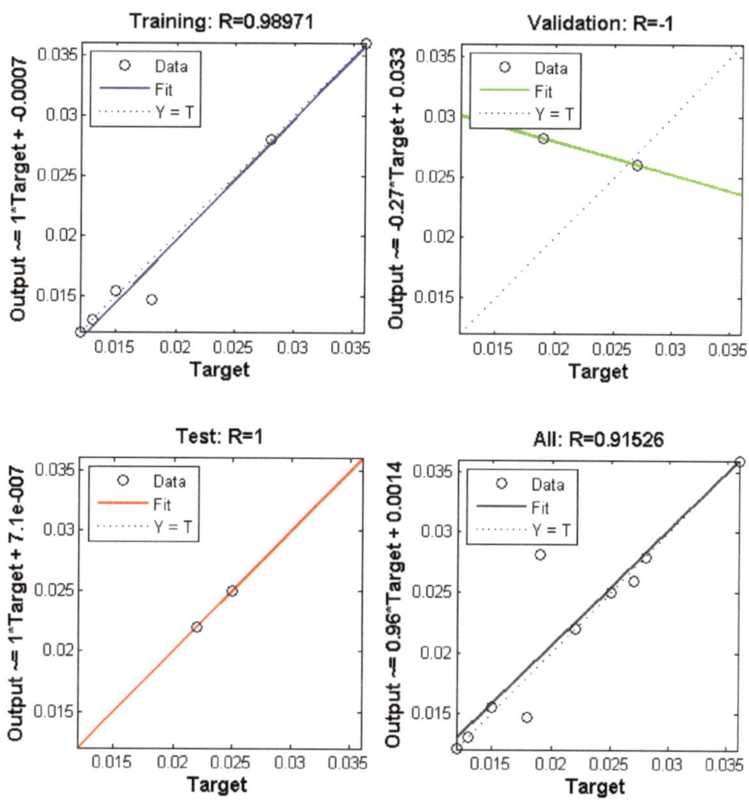

Figure 36.
Results of multilayer artificial neural network (MANN) training.

groups: one that met the quality requirements and one that failed to meet the quality requirements. (b) The result is linearly classified into three groups as low, medium, and high rotation speed range. In low and high rang, the machining quality cannot meet the quality requirement. Sometimes, the machining quality classification is not linear as shown in (c). The classification quality is not only based on deep learning technology but also depends on the quantity of parameters and the machining material. The matching quality can be further classified into grades of quality such as grade A, B, and C, as shown in **Figure 36**, based on machining tolerances.

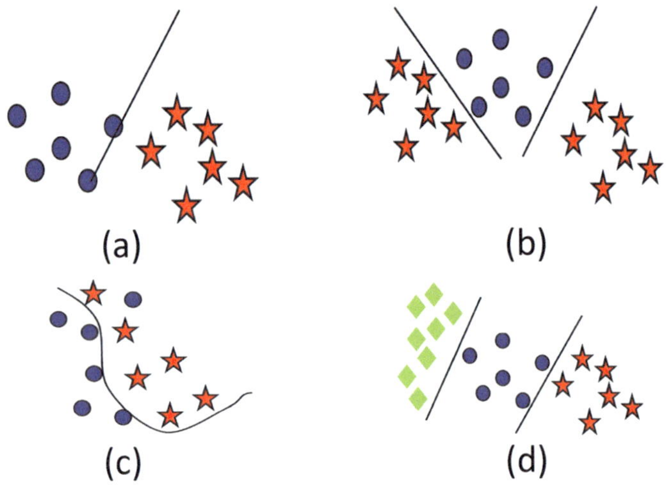

Figure 37.
Data classification results based on deep learning. (a) Two classes linear classification, (b) multi-classes classification, (c) two classes no-linear classification, and (d) multi-classes no-linear classification.

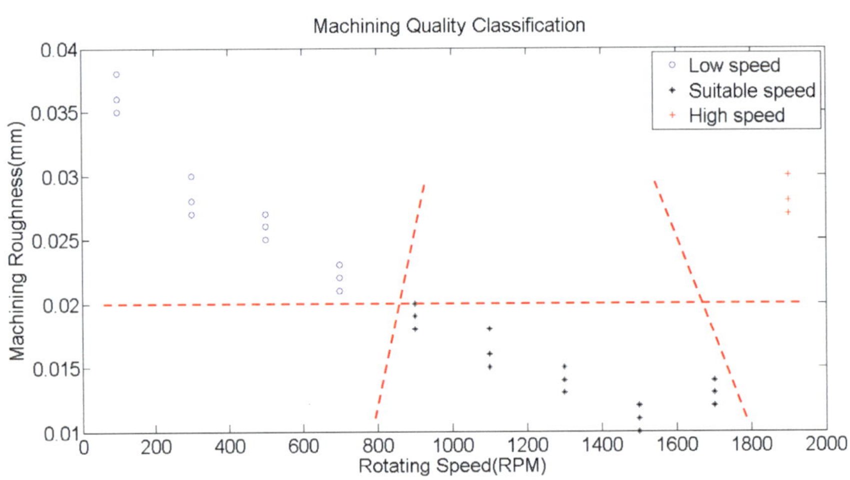

Figure 38.
Classification results of machining nylon with varied speed of CNC.

Figure 38 shows classification results of machining Nylon with variable speed of CNC based on deep learning. If the machining roughness threshold is set as 0.02 mm, the spindle speed of CNC should be more than 900 rpm and less than 1800 rpm.

6. Conclusions

This chapter presents our research studies on a parameter optimization method of product processing procedure for CNC machine tool with three common systematic errors compensation approaches.

In Section 1, we introduced mechanism optimization, energy consumption optimization, and processing optimization of a CNC machine tool. Mechanism optimization is mainly used to optimize the CNC machine tool to reduce machine systematic errors. On the other hand, due to huge electrical energy consumption of CNC machine tool under operational mode, energy consumption optimization should be considered. There are tremendous researches on the first two aspects but less on the last one. Therefore, in the next section, our studies focus on the processing optimization approach.

In Section 2, we briefly introduced the working principle of CNC machine tools. In order to reduce machining error, the cutting force has to be monitored constantly. The machining of the complex surface can be decoupled into simpler ones for path optimization in order to minimize speed fluctuation-induced machining error. For the same purpose, machining tool overcutting compensation and two major indicators S_t and S_q of the machining accuracy are introduced and an optimized profile is proposed.

In Section 3, we studied tool deflection issue during machining. The merits and drawback of two commonly used displacement sensors, capacitive probe and eddy current gap sensor, are compared and recommended for different scenarios. In this section, 5G AI edge computing technology for signal pre-processing and genetic programming for control algorithm parameters optimization are proposed. The experimental results with tool deflection compensation are reduced by 40%.

In Section 4, we studied error feedback compensation approach. The CNC and servo actuator are in the close loop for error feedback compensation. A bandpass filter is adopted to reduce the noise in order to reconstruct the control signal. The experimental result is reduced by 60%.

In Section 5, we studied processing parameter optimization. The cloud-based intelligent network configuration for processing parameter optimization is proposed. In this section, the method is mainly based on deep learning approach such as GP, MANN, and SVM to identify the product processing optimized parameters. Before deep learning approach is employed in cloud computing with expert system and database, all the information in the train data instance such as raw material type and dimensions, CNC machine tool systematic parameters, and machining operational conditions and vibration outputs of the CNC machine tool must be sent to a cloud-based computing center. The amplitudes and frequencies of the CNC machine tool outputs are determined by the waterfall of FFT after the signal is preprocessed through 5G AI edge computation. The functionality and performance of the proposed technique are verified with simulations as well as with the experimental data and results, which proves the effectiveness of the proposed scheme.

Acknowledgements

First and foremost, I would like to thank East China Jiao Tong University for providing the research funding (No: 2003419018) for this study. I would like to express my gratitude to Dr. Yang Yang, Director of Engineering Experiment Teaching Center of East China Jiao Tong University, for his invaluable support, Mr. wang Guorong and Ms. Zhang han for their meaningful experimental setup in this study.

I would like to thank for all the help I received from my colleagues, Dr. Liu Yande, Dr. Hu Guoliang, Dr. Zhou Jianming, Dr. Chen Qiping and Dr. Tu Wenbing, from School of Mechatronics & Vehicle Engineering in East China Jiao Tong University, who helped me one way or another.

Finally, I would like to thank my son, Mr. Yu Shi Jie, for improving the writing of this document and my family for their unwavering support.

Author details

YinQuan Yu
East China Jiao Tong University, Nan Chang, Jiang Xi Province, China

*Address all correspondence to: yu_yinquan@ecjtu.edu.cn

References

[1] Shukun C, Heng Z, Xiangbo Z, Qiujuan Y, Changsheng A. Software and hardware platform design for open-CNC system. In: 2008 IEEE International Symposium on Knowledge Acquisition and Modeling Workshop; 21–22 December 2008; Wuhan. New York: IEEE; 2008. pp. 139-142. DOI: 10.1109/KAMW.2008.4810444

[2] Prado Y, Valiño G, Blanco D, Suárez CM, Álvarez BJ. Models for stiffness characterization of the spindle-chuck system in a CNC lathe for prediction of deflections in CAPP. In: 2010 IEEE 15th Conference on Emerging Technologies & Factory Automation (ETFA 2010); 13–16 September 2010; Bilbao. New York: IEEE; 2010. pp. 1-7. DOI: 10.1109/ETFA.2010.5641361

[3] Yang Y et al. Geometric error modeling and compensation of the dual-driving feed worktable. In: 2019 6th International Conference on Systems and Informatics (ICSAI); 2–4 November 2019; Shanghai. New York: IEEE; 2020. pp. 86-91. DOI: 10.1109/ICSAI48974.2019.9010391

[4] Gang L, He X. Analysis on the basic application technology of CNC machine tools. China New Technology & Products. 2019;**23**:94-95. DOI: 10.1590/S0100-73862002000300009

[5] Chen G, Liang Y, Sun Y, Chen W, Wang B. Volumetric error modeling and sensitivity analysis for designing a five-axis ultra-precision machine tool. International Journal of Advanced Manufacturing Technology. 2013;**68**: 2525-2534. DOI: 10.1007/s00170-013-4874-4

[6] Gabriel S, Raul C. Gouveia R. designing a novel feeding system for CNC turning machines. Procedia Manufacturing. 2018;**17**:1144-1153. DOI: 10.1016/j.promfg.2018.10.020

[7] Lin Z, Guo R, Chen L, Geng C, Wang F. Design and development of the self-adaptive tool path decision-making CNC platform. In: 2012 International Conference on Biomedical Engineering and Biotechnology; 28–30 May 2012; Macao. New York: IEEE; 2012. pp. 135-137. DOI: 10.1109/iCBEB.2012.126

[8] Lapsomthop O, Wongsirirax N, Kititeerakol A, Sawangsri W. Design and experimental investigation on 3-component force sensor in mini CNC milling machine. Materials Today: Proceedings. 2019;**17**(4):1931-1938. DOI: 10.1016/j.matpr.2019.06.232

[9] Moreira LC, Li WD, Lu X, Fitzpatrick ME. Supervision controller for real-time surface quality assurance in CNC machining using artificial intelligence. Computers and Industrial Engineering. 2019;**127**:158-168. DOI: 10.1016/j.cie.2018.12.016

[10] Cheng Q, Zhao H, Zhao Y, et al. Machining accuracy reliability analysis of multi-axis machine tool based on Monte Carlo simulation. Journal of Intelligent Manufacturing. 2018;**29**: 191-209. DOI: 10.1007/s10845-015-1101-1

[11] Xie C, Zhang WM, He XY. Kinematic analysis and post-processing algorithm research for 5-axis CNC machine tools with a universal head. In: 2009 IEEE International Conference on Industrial Engineering and Engineering Management; 8–11 December 2009; Hong Kong. New York: IEEE; 2009. pp. 2309-2313. DOI: 10.1109/IEEM.2009.5373031

[12] Cortina M, Arrizubieta JI, Ruiz JE, Ukar E, Lamikiz A. Latest developments in industrial hybrid machine tools that combine additive and subtractive operations. Materials (Basel). 2018; **11**(12):2583. DOI: 10.3390/ma11122583

[13] Wang K-C. Optimal high-rigidity structure design for CNC machine tools using CAE technique. Engineering Computations. 2014;**31**(8):1761-1777. DOI: 10.1108/EC-11-2012-0296

[14] Yang J, Mayer JRR, Altintas Y. A position independent geometric errors identification and correction method for five-axis serial machines based on screw theory. International Journal of Machine Tools and Manufacture. 2015; **95**:52-66. DOI: 10.1016/j.ijmachtools. 2015.04.011

[15] Zhu X, Bao J, Wang J, Chen F, Li X, Zhang X. A comprehensive maintainability evaluation methods for subsystems of CNC machine tools. Journal of Physics: Conference Series. 2018;**1074**:012144. DOI: 10.1088/1742-6596/1074/1/012144

[16] Zhaojun Y, Dong Z, Chuanhai C, et al. Reliability modelling of CNC machine tools based on the improved maximum likelihood estimation method. Mathematical Problems in Engineering. 2018;**2018**:1-11. DOI: 10.1155/2018/4260508

[17] Li C, Zhang X, Zhang Q, Li H. Numerical simulation analysis of temperature field for motorized spindle of high-grade CNC machine tool based on ANSYS. Key Engineering Materials. 2010;**455**:33-36. DOI: 10.4028/www.scientific.net/KEM.455.33

[18] Vicente García J. Development of Valid Models for Structural Dynamic Analysis [thesis]. South Kensington, London, United Kingdom: Imperial College London; 2008

[19] Bonte MHA, Fourment L, Do T, et al. Optimization of forging processes using finite element simulations. Structural and Multidisciplinary Optimization. 2010;**42**:797-810. DOI: 10.1007/s00158-010-0545-3

[20] Erfei L, Xinghua N, Yiguang S, Guo Y. Study on the reason of static stiffness of the whole machine in non-circular phenomenon when vertical machining center milling circle. IOP Conference Series: Materials Science and Engineering. 2018;**452**:042038. DOI: 10.1088/1757-899X/452/4/042038. 15–16 September 2018; Melbourne

[21] Bosetti P, Bruschi S. Enhancing positioning accuracy of CNC machine tools by means of direct measurement of deformation. International Journal of Advanced Manufacturing Technology. 2012;**58**:5-8. DOI: 10.1007/s00170-011-3411-6

[22] Kumar V, Kumar BJK, Rudresha N. Optimization of machining parameters in CNC turning of stainless steel (EN19) by Taguchi's orthogonal array experiments. Materials Today: Proceedings. 2018;**5**:11395-11407. DOI: 10.1016/j.matpr.2018.02.107

[23] Shaoke W, Xiaoliang J, Kumar MN, Jun H. Effect of vibration assistance on chatter stability in milling. International Journal of Machine Tools and Manufacture. 2019;**145**:103432. DOI: 10.1016/j.ijmachtools.2019.103432

[24] Lee K, Lee T, Yang M. Tool wear monitoring system for CNC end milling using a hybrid approach to cutting force regulation. International Journal of Advanced Manufacturing Technology. 2007;**32**:8-17. DOI: 10.1007/s00170-005-0350-0

[25] Zhang C, Guo S. Error compensation model considering tool wear in milling difficult to cut materials. International Journal of Computer Integrated Manufacturing. 2017;**30**(8): 822-838. DOI: 10.1080/0951192X.2016.1210232

[26] Andrew W. An Early Warning Monitoring System for CNC Spindle Bearing Failure [thesis]. Tigerprints: Clemson, South Carolina, United States: Clemson University; 2011

[27] Oral A, Cakir MC. Automated cutting tool selection and cutting tool sequence optimisation for rotational parts. Robotics and Computer-Integrated Manufacturing. 2004;**20**: 127-141. DOI: 10.1016/j.rcim.2003.10.006

[28] Han J, Wu L, Yuan B, Tian X, Xia L. A novel gear machining CNC design and experimental research. The International Journal of Advanced Manufacturing Technology. 2017;**88**:5-8. DOI: 10.1007/s00170-016-8883-y

[29] Yang X-Y, Tang J-Y. Research on manufacturing method of CNC plunge milling for spur face-gear. Journal of Materials Processing Technology. 2014; **214**:3013-3019. DOI: 10.1016/j.jmatprotec.2014.07.010

[30] Xing Y, Wang T. Accuracy enhancement in manufacture of spiral bevel gear with multi-axis CNC machine tools by a new compensation method. In: 2011 International Conference on Consumer Electronics, Communications and Networks (CECNet); 11–13 March 2005; Xianning. New York: IEEE; 2011. pp. 3891-3894. DOI: 10.1109/CECNET.2011.5768862

[31] Lee S-Y. In-process tool condition monitoring systems in CNC turning operations [Thesis]. Ames, Iowa, United States: Iowa State University; 2006

[32] Smith GC. Using microwave Doppler radar in automated manufacturing applications [Thesis]. Ames, Iowa, United States: Iowa State University; 2004

[33] Renn JC, Hsu WJ, Liao WC. Energy efficient lathe turret design using load sensing control scheme. In: 2014 International Symposium on Computer, Consumer and Control. 10–12 June 2014, Taiwan. New York: IEEE; 2014. pp. 51-54. DOI: 10.1109/IS3C.2014.26

[34] Altıntaş RS, Kahya M, Ünver HÖ. Modelling and optimization of energy consumption for feature based milling. International Journal of Advanced Manufacturing Technology. 2016;**86**: 3345-3363. DOI: 10.1007/s00170-016-8441-7

[35] Guoyong Z, Chunhong H, Jianfang Q, Cheng X. Energy consumption characteristics evaluation method in turning. Advances in Mechanical Engineering. 2016;**8**(11):1-8. DOI: 10.1177/1687814016680737

[36] Lv L, Deng Z, Liu T,Wan L, Huang W, Yin H, et al. A composite evaluation model of sustainable manufacturing in machining process for typical machine tools. Processes. 2019;7 (2):110. DOI: 10.3390/pr7020110

[37] Liu ZJ, Sun DP, Lin CX, Zhao XQ, Yang Y. Multi-objective optimization of the operating conditions in a cutting process based on low carbon emission costs. Journal of Cleaner Production. 2016;**124**:266-275. DOI: 10.1016/j.jclepro.2016.02.087

[38] Jianmai S, Zhong L, Luohao T, Jian X. Multi-objective optimization for a closed-loop network design problem using an improved genetic algorithm. Applied Mathematical Modelling. 2017; **45**:14-30. DOI: 10.1016/j.apm.2016.11.004

[39] Böllinghaus T et al. Manufacturing Engineering. Springer Handbook of Mechanical Engineering. Berlin/Heidelberg: Springer; 2009. pp. 523-785

[40] Makhanov SS. Adaptable geometric patterns for five-axis machining: A survey. International Journal of Advanced Manufacturing Technology. 2010;**47**:1167-1208. DOI: 10.1007/s00170-009-2244-z

[41] Wasif M, Iqbal SA, Ahmed A, et al. Optimization of simplified grinding wheel geometry for the accurate

generation of end-mill cutters using the five-axis CNC grinding process. International Journal of Advanced Manufacturing Technology. 2019;**105**: 4325-4344. DOI: 10.1007/s00170-019-04547-8

[42] Mou W, Jiang Z, Zhu S. A study of tool tipping monitoring for titanium milling based on cutting vibration. The International Journal of Advanced Manufacturing Technology. 2019;**104**: 3457-3471. DOI: 10.1007/s00170-019-04059-5

[43] Umbrello D, Jawahir IS. Numerical modeling of the influence of process parameters and workpiece hardness on white layer formation in AISI 52100 steel. International Journal of Advanced Manufacturing Technology. 2009;**44**: 955-968. DOI: 10.1007/s00170-008-1911-9

[44] Reyes Uquillas DA, Yeh S. Tool holder sensor design for measuring the cutting force in CNC turning machines. In: 2015 IEEE International Conference on Advanced Intelligent Mechatronics (AIM); 7–11 July 2015; Busan. New York: IEEE; 2015. pp. 1218-1223. DOI: 10.1109/AIM.2015.7222705

[45] Fountas NA, Vaxevanidis NM, Stergiou CI, et al. A virus-evolutionary multi-objective intelligent tool path optimization methodology for 5-axis sculptured surface CNC machining. Engineering with Computers. 2017;**33**: 375-391. DOI: 10.1007/s00366-016-0479-5

[46] Du X, Sun Y. Dynamic transmission error analysis for CNC machine tools under variable speed condition. In: 2018 IEEE 4th Information Technology and Mechatronics Engineering Conference (ITOEC), 14–16 December 2018; Chongqing. New York: IEEE; 2018. pp. 113-120. DOI: 10.1109/ITOEC.2018.8740694

[47] Patil RS, Jadhav SM. Boring parameters optimization for minimum surface roughness using CNC boring machine with passive damping material. In: 2017 2nd International Conference for Convergence in Technology (I2CT), 7–9 April 2017; Mumbai. New York: IEEE; 2017. pp. 300-303. DOI: 10.1109/I2CT.2017.8226140

[48] Nie X. Application of neural network for thermal error compensation in CNC machine tool. In: 2011 IEEE 2nd International Conference on Computing, Control and Industrial Engineering, 20–21 August 2011; Wuhan. New York: IEEE; 2011. pp. 211-215. DOI: 10.1109/CCIENG.2011.6007995

[49] Bobyr MV, Abduldaiem AN, Abduljabbar MA. Cooled cutter control algorithm based on fuzzy logic. In: 2017 International Conference on Industrial Engineering, Applications and Manufacturing (ICIEAM), 16–19 May 2017; St. Petersburg. New York: IEEE; 2017. pp. 1-5. DOI: 10.1109/ICIEAM.2017.8076168

[50] Liu J et al. Thermal boundary condition optimization of ball screw feed drive system based on response surface analysis. Mechanical Systems and Signal Processing. 2019;**121**:471-495. DOI: 10.1016/j.ymssp.2018.11.042

[51] Jie G, Agapiou JS, Kurgin S. Global offset compensation for CNC machine tools based on Workpiece errors. Procedia Manufacturing. 2016;**5**: 442-454. DOI: 10.1016/j.promfg.2016.08.037

[52] Béres M, Paripás B. Measurements of vibration by laser doppler method in the course of drilling. In: Jármai K, Bolló B, editors. Vehicle and Automotive Engineering. Lecture Notes in Mechanical Engineering. Vol. 2. Cham: Springer; 2018. DOI: 10.1007/978-3-319-75677-6_16

[53] Li X et al. Monocular-vision-based contouring error detection and compensation for CNC machine tools. Precision Engineering. 2019;**55**:447-463. DOI: 10.1016/j.precisioneng.2018.10.015

Chapter 3

Interlinking Industry 4.0 and Academia through Robotics and Automation: An Indian Perspective

Prajwal Prabhu, Subin Raj and Abhra Roy Chowdhury

Abstract

Robots and automation systems are growing rapidly in the society globally with an annual global sales value of 16.5 billion USD in 2018 according to the International Federation of Robotics (IFR). They have found a commonplace not only in industries and service sector but also in households. This has attracted heavy investment by industries globally in the research and development of robotics and its applications. Understanding this rising trend in the industry and society, there is an obvious need for expertise and future workforce in robotics. According to the IFR report, India has recorded a growth of 39% in 2018 compared to the previous year in terms of annual robot installations. The annual report of the Confederation of Indian Industry (CII) in 2019 recommends robotics and automation as one of the prime areas of focus toward the development of national policies on Industry 4.0. One such interlinking initiative in robotics research and innovation has started at the Centre for Product Design and Manufacturing (CPDM) in the Indian Institute of Science (IISc). The project is designated under India's first Industry 4.0-compliant Smart Factory R&D platform in a unique academic set-up. It aligns with the policies of Govt. of India to boost vision Industry 4.0 for India's technological and economic transformation.

Keywords: Industry 4.0, robotics, automation, autonomous systems

1. Introduction

With the advent of industrial development, humans have witnessed its four phases starting from mechanized processes, mass production using electricity, assembly lines and now finally culminating to Industry 4.0. It is also commonly referred as Smart factory where the physical processes of factory are monitored by decentralized cyber-physical systems to take autonomous and individual decisions. The physical systems i.e. machines are configured with the Internet of Things (IoT), communicating in real time to each other and with human operators over the web representing another vital component called interoperability. Indian Institute of Science (IISc) has been developing a "Smart Factory called I4.0India@IISc" that will act as a Common Engineering Facility Centre (CEFC) at the Centre for Product Design and Manufacturing (CPDM), under the SAMARTH Udyog Bharat 4.0 program of the Department of Heavy Industries (DHI), GoI in [1].

The CEFC builds on the earlier work at IISc that initiated India's first smart factory R&D platform with seed collaboration fund from the various multinational and national companies worldwide. The CEFC intends to develop two contrasting factory platforms. One is highly automated, with data-intensive processing machines such as 5-axis CNC, metal additive manufacturing, metal laser routers etc. that are handled by Industrial robots, collaborative robots, automated guided vehicles, etc. The other platform is highly manual, with legacy machines with data layer, where handling is also carried out manually. The distinction is that all elements in both the Platforms (people, process, parts, tools and environment) are to be connected via IoT-based communication networks and data are stored and analytics carried out using cloud computing. The intent is to use the platforms as demonstration centers in which factories of the future will be explored to see their potential and challenges, and use the platforms for advancing research, innovation, training and support for practice. A major aim of this CEFC is to support Industry 4.0 in MSMEs of India by developing "advanced yet affordable" technologies and solutions for their use. The CEFC is supported by Tata Consultancy Services (TCS), Yaskawa, Faurecia, Toyota Kirloskar Motors (TKM) and Ashok Leyland as industrial partners who contribute with fully funded research projects. The CEFC is also supported by faculty members drawn from 12 departments of IISc as collaborators. A variety of research-intensive technologies are being developed at the CEFC for the benefit of industry. The technologies under development include smart resource bins for waste tracking, smart inspection systems, AR/VR-based digital twins and digital walkthrough systems, 3D printing systems for printing devices with embedded electronics, PCB assembly support systems, multi-modal robots, affordable door localization systems, etc. The CEFC is meant to serve the following purposes:

- Use of the factory with its two platforms as demonstrators for Industry 4.0 (development of platform should be completed in next 6 months)

- Utilization of the platforms for trying out new building blocks, technologies and solutions for their potential effects on KPI before committing investment on these.

- Playing with various combinations of the two platforms for creating different variants of factories of the future and seeing their impact on KPIs.

- Using the factory as a springboard for supporting development of new innovations and start-ups for Industry 4.0 (1 start-up initiated, 10 existing start-ups already supported as planned)

- Creating Indigenous R&D capabilities in the area for India's Industry 4.0 by training researchers and developing deep tech knowledge. (11 PhD students and 20 researchers are being trained, over 20 technologies, tools and solutions are being developed with industrial significance)

- Providing awareness, training and education, especially high-end training to develop future leaders in manufacturing (over 500 people already trained, which is 2.5 times more than originally planned; a new MTech and PhD program in manufacturing initiated at IISc, an international conference on Industry 4.0 and Advanced Manufacturing I-4AM'19 initiated)

In continuation to this we have identified the opportunities and challenges for human robot collaboration in Industry 4.0 set-up. This is mentioned in subsequent sections.

2. Human factors in logistics and warehouse industry

A supply chain in a manufacturing environment involves manufacturers, suppliers, and other logistical entities that would entail dealing with the customers' requests. Modern management practices are designed to reduce the inventories; thus, warehouses play a crucial role in customer services and product availability. Logistical support provided to improve the warehouse order picking scenario would entail reduction of time in retrieving of the object and its successful shipment for delivery. Customer satisfaction in such a scenario would require for an efficient and flexible warehouse order handling processes. However, any small changes in the customer end of the supply chain would create a large ripple in the stock levels and orders along the supply chain also called as a "Bullwhip Effect." Increasing demand at the nth tier in a supply chain will be over-compensated by an even greater increase of the demand at the n+1th tier of the chain in [2]. As a result, the curve mapping the stock levels along the time looks similar to a whiplash, hence the name "Bullwhip Effect." Some of the key aspects of the human factors in the logistics industry are quantified as Perceptual Factor, Mental Factor, Physical Factor, Psychological Factor in [3]:

1. Perceptual Factor: Perception to new things will play a role to changing needs of warehousing environment. This will play a very important factor in determining the overall productivity of the warehouse. This factor will amount to 24.4% in the order picking process. Some of the key things that will affect in this factor would include information technology, ability to read, confusion created due to negligence and ability to accept new technology. It is of the utmost importance that the laborers are skilled to understand the upcoming technology that will reduce the overall burden of their work.

2. Mental factor: The mental factor in any order picking process will be of maximum importance and will amount to 45.2% in the order picking process. Some of the key aspects in this factor would be as follows: cognitive ability, ability to learn new technology will be a very important factor to ensure that the order picking process can be done in a very efficient way and will amount to 37.3%. Some of the other factors include behavior, training. To ensure the productivity of the warehouse, the laborers should have decent working environment where they are more receptive and learn modern technologies and new ideas that will improve the efficiency of the system.

3. Physical Factor: The physical factor in order picking process will amount to 16.3%. Some of the key aspects that will affect the order picking process are: Having an improper workload will lead to musculoskeletal diseases in the human beings causing back pain and fatigue. Further improper material handling may lead to material damages that will increase the overall cost in the production. It is of utmost importance that the workload of all laborers should be uniformly distributed, to avoid the breakdown of the warehouse machinery.

4. Psychological Factor: Factors like work environment, stress, monetary incentives also play a very important role in ensuring that human beings are motivated to do quality work and ensure that the order handling process runs in a very smooth and efficient way. Laborers should have proper incentives and can be in the form of monetary or otherwise, to ensure that they are motivated throughout to ensure that the productivity of the warehouse is not dropped.

3. Indian scenario for the human factors in logistical/warehousing industries

Logistical cost in Indian scenario accounts to 13–17% of the overall GDP, which is nearly double the GDP of the developed countries like US, France, UK, etc. in [4]. Main reasons for such high costs are due to lack of efficient transport systems. Further it is seen that there is a limited focus on robotics and automation which would reduce the majority of the human works. The main percentage of the logistical cost in India is borne by the end users. The future trends in warehousing industry would involve companies consolidating large warehouse spaces and developing futuristic technologies that would improve the warehousing operations leading to better and efficient productivity. A huge amount of flexibility can be incorporated by including human operators in the warehousing industry. Human being can react quickly to any operational changes or urgent requirements of the orders and maneuver in tight spaces that lead to efficiency in the order picking systems. Some of the key measures taken by the Indian government to improve the logistical scenario in India include the Creation of exclusive Warehousing Promotion Zones. Create logistics and warehousing policies separate from industrial promotion policies, Allocation of areas exclusively for warehousing within Industrial Parks, Setting up of Skill Augmentation Centers in [4]:

1. Creation of exclusive Warehousing Promotion Zones: To create modern warehouses and facilitate modern warehousing, need to create exclusive warehousing zones is stressed to develop modern warehousing infrastructures.

2. Create logistics and warehousing policies separate from industrial promotion policies: State governments are mandated to form separate policies for development of the logistics and warehousing industry so that the industries can grow to its full potential.

3. Allocation of areas exclusively for warehousing within Industrial Parks: Adequate storage facilities for raw materials and finished goods need to be provided in the industrial parks where the manufacturing facilities are provided.

4. Setting up of Skill Augmentation Centers: It is of the utmost importance to set up a skill augmentation center so that warehouses can be modernized to improve productivity and efficiency. Some of the modern technologies that can be included in the warehousing include using of cloud to access the warehouse inventories, deployment of drones in the warehousing industry. Using of IoT and automation can improve the overall performance of the system.

4. Need for autonomous mobile robot in manufacturing industries

Automation in the manufacturing industry will ensure that the end result will be achieved in a more effective way thereby reducing the human effort. Automated guided vehicles are the most extensively used robots in the industry; however, these lack the self-regulating mobility and need navigation lines along the shop floor. Further material handling while using the AGV will require human effort which in some cases will reduce the productivity. The need for human operators to undertake planning activities and the lack of real time feedback of the shop floor's inventory level will make the efficient flow of materials tough.

Autonomous mobile robots are machines designed to maneuver itself in an obstacle filled environment through the use of sensors and feedback. Due to its design it can be used in manufacturing industry thereby reducing wear and tear in different industrial components and improve the productivity. AMR's provide a wide range of advantages such as: reduced floor traffic, flexibility, reliability, self-regulating mobility etc. in [5].

The need for using the autonomous mobile robots in manufacturing environment is as follows:

1. To increase productivity and efficiency

2. To improve material handling

3. To improve the safety of human operators in high risk, stress environments

4. To perform repetitive tasks so that the human beings can concentrate on cognitive and creative tasks.

5. To enhance overall revenue by delivering order in perfect fulfillment rates and better customer satisfaction.

6. To reduce the factory traffic by navigating in tight and complex environments.

5. Human robot interactions in logistics/warehousing industry

Human machine interaction is one of the most crucial aspects of the warehousing industry. Since there is a collaboration between human and machine in the warehousing industry, efficient interaction between the two systems would ensure that the process runs in a smooth way in [6]. In an HMI system, machines relieve human from performing repetitive tasks, and human can focus on working in creative and cognitive tasks. Thereby such collaborations would improve the efficiency and reliability of the overall system. Human beings will not be able handle heavy objects in the warehousing industry. Such things will lead to the musculoskeletal damage in human beings and factor into improper material handling. The advent automation has reduced the burden on the human in the warehousing industries. However, there is a reduction of skill in the warehouse workers due to the increased workload on the human beings in [6]. Thereby creating a concept of adaptive automation where there is a continuous interaction between the human and machine considering the ever-changing needs of a logistics/warehousing setup.

Figure 1 shows the possible types of interaction in human machine system, which is adapted from "Formalizing Human–Machine Interactions for Adaptive Automation in Smart Manufacturing" [6].

In general, the possible human interactions that are seen in a warehousing environment can be classified as Human machine interaction, Human task interaction, Supervisor system interaction in [6].

Typical human machine interaction system would consist of human supervisor, several manufacturing cells that and each cell consists of human operator, machine and a human machine interface. The whole system can be viewed as a cell layer and supervisor layer. The interfaces at the cell layer are associated with the UIs for human operators who manipulate the machines to achieve the logistical goals and respond to unexpected situations either by reporting unanticipated events to the supervisor or performing the exception handling tasks issued by the supervisor.

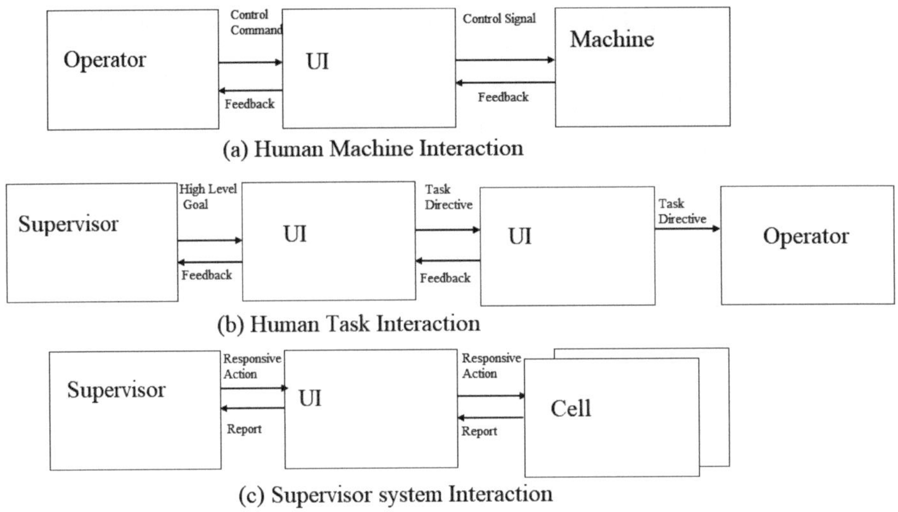

Figure 1.
Description for a human-machine system based on the type of interaction.

Information about machine state is delivered to the human operator through the interface and the operator controls the machine directly through his commands and suitably maneuvers the machine. The operator will notify the supervisor about any interrupts that may affect the overall performance of the system thereby allowing the supervisor to perform exception handling tasks. The interfaces at the supervisor layer are for a human supervisor in the human machine system. In normal scenarios the supervisor will monitor the key monitoring indices for logistical warehouses and maintain the task progress. In cases of exceptions the supervisor has to advice the operator to take the corrective actions. Exception handling task supportability is associated with responsive management concerning the uncertainty inherent in almost every logistical system. Designing of Human machine systems should take into account that human machine system has to assist humans or supervisor in coping with an unforeseen scenario. Due to a high degree of complexity, the problems of exceptions only by the means of autonomous robots will be a challenge, thereby creating the need for human involvement.

6. Present problems in autonomous navigation in factory set-up

Due to the complex nature of the AMR's, some of the key limiting factors that decide the acceptance of autonomous mobile robots in the manufacturing industry are in [4]:

1. Self-efficiency: It decides the ability of the user to adopt to new technology and perform given tasks under different circumstances. Some of the sub-factors include understanding the usage and application of AMR, ability to use the AMR without any assistance. It is highly necessary that human operator is well educated about the application of the AMR's so that upon interrupt, necessity suitable changes can be incorporated to the systems without bringing the whole production to a standstill.

2. Performance expectancy: Since the AMR comprises of many sensitive electronic components, the reliability of the overall system and the performance

of the said system will dictate the productivity in the usage of AMR's in the industry. Further due to the lack of regular maintenance schedule will bring down the performance expectancy of the robot.

3. Perceived risks: AMR's are prone to a traffic collision due to hardware failure, software failure which will often lead to improper material handling, thereby creating a scene of uncertainty towards performance and productivity. Such incidents usually end with a financial loss to the manufacturers.

4. Intentions to use: Intention of the use of AMR's is governed by users and the end application. If the technology is used as per the purpose of its creation then it will lead to easing out of work, thereby boosting the industries. Due to the software functionality of the AMR's once if the steps involved in the manufacturing process is trained, it will retain these steps thereby reducing the intensity of the work.

5. Privacy: Breach of private information through hacking of the software used will lead to wrong marketing, scrutiny, etc. vehicle to vehicle communication will involve transferring of vehicular data in order to have better productivity but however leads to breach of privacy. So, privacy of data will be important in the autonomous environment to ensure that the data generated by the systems will not get into the wrong hands.

6. Reliability: The reliability of the AMR's will mainly depend on the following factors: precision, accuracy, safety, security. Some of the sub-factors include getting confused due to extenuating circumstances of the everchanging needs of the warehouse, precision and the accuracy with which the system responds to the changing needs of a manufacturing industry, mean time between failure.

Further there are other constraints that limit the overall performance of the AMR's in the factory setup in [7]:

1. Getting the software correct so that the robot can navigate the factory setup autonomously and do the complex tasks in an efficient way.

2. Gathering enough real-world data so that the robot can accommodate the real-world changes in the factory setup.

3. Creating precise motion control so that the robot can maneuver in tight spaces and material handling is done in an efficient way.

4. Reducing the false positives that may occur by training the robot in enough real-world scenarios and distinguishing the data obtained between the actual data and the false positive ones.

5. Present Solutions Used for Autonomous Navigation in Factory Set-Up.

One of the evident problems that may be seen while using an AMR is the process of motion planning for an unstructured environment like a warehouse where there are continuous upgrades of the warehouse technology and thereby will be constant updating of the maps. In such a scenario where the start and the finish points are fixed, the motion planning will be complex. There are infinite possibilities for achieving the trajectory for a given start and finish. The best way would be to use a probabilistic approach to generate the trajectory using the concept of least action. The concept of least action will use the simple fact that the best possible path out of the infinite possible paths from a start point to the finish point would be the one that consumes the least energy in [8]. The principle of least action will generate a best possible path out of the infinite possibilities that are available from starting point to finish point considering all of the infinite paths that are possible. While using the concept of least action all of the possible paths are discretized for a small interval that is decided by the user, and the trajectory generated for this discretized instant for all the paths will be a straight line as shown in the **Figure 2** that is selected in [9].

To find the energy of each of the infinite path that is essential in the computation of the least action, there will be a Lagrangian that is defined for all the paths that will be a function of time in [8]. Further an action function is defined, which will be an integral of the Lagrangian for the path from start to the finish.as per the Eq. 1. The main objective would be to minimize this function, so that the best optimal path is generated.

$$S = \int_{ti}^{tf} L\, dT \tag{1}$$

where S = the action function that is defined for a given start and finish.

L = Lagrangian that is defined over the path which is the difference of the kinetic and the potential energy and is the function of the time.

ti = start time.

tf = finish time.

The Lagrangian for the system is defined as

$$L(x,\dot{x}) = K(\dot{x}) - P(x) \tag{2}$$

where $k(\dot{x})$ = overall Kinetic energy for the path.

P(x) = overall Potential energy for the path.

The physical significance of the action function would be that, the function would eliminate the infinite path which is not practical and will keep only the paths that make sense, from where the optimal path is selected. Further by discretizing the path for small instants can be useful for the process of real time obstacle detection in the environment and generating a trajectory that can be obstacle free. The path taken by the system that is generated through the concept of least action will give the complete history of the path that is taken to achieve the required goal which means that path taken by the robot is dependent on the path that the robot has taken

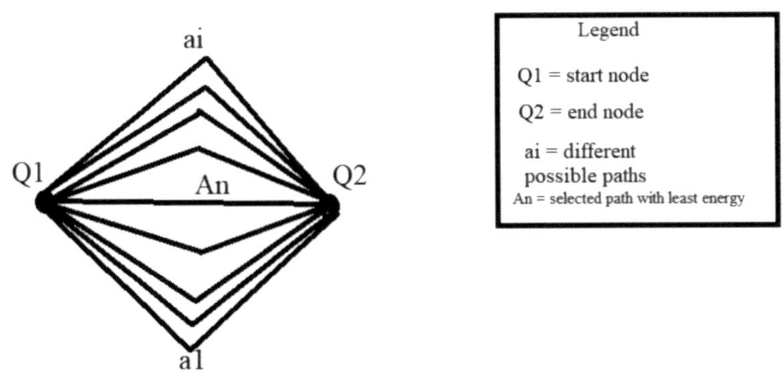

Figure 2.
Infinite possible paths that are possible for a given start and finish.

previously. The simplicity of this approach makes it more useful in using the proposed method in complex environments. So, the institute looks at incorporating the concept of least action for the motion planning of autonomous mobile robots that is developed at the common engineering facility center. The technology developed at the common engineering facility center is discussed in the paragraphs below.

The present solutions those are developed by the industry keeping in mind the different problems like the software complexity, performance, flexibility in the harsh conditions of a warehouse environment are as follows in [10, 11]:

1. Goods to person picking robots: companies like IAM, robotics, gray orange and blue offer mobile robot solutions that are loaded with modern technology and add new efficiency to the tasks done in factory setup. These robots can be programmed to have flexible routes in the factory set-up and requires less infrastructure to set up. The computer interface is friendly and easy to understand. Thereby the human machine interaction happens in a smooth way.

2. Self-driving forklift: Linde's new automated forklift consists of navigation lasers, 3D camera and visual and acoustic warning indicator that enables to move safely around human workers. The company also includes the feature of detection of real-time obstacles and adjust suitably. These things can work in fleet thereby optimizing the performance of the warehouses. These are being used in Kerala, Gujarat, Madhya Pradesh for different industries like electronics, consumer. The system further is mainly used as an artificial intelligence solution to the said product and has been installed in the manufacturing setup at ITC.

3. Autonomous inventory robots: FETCH's tag surveyor consists of three RFID mounted for optimal coverage and has reliability and consistently detect the products up to 25 feet away, thereby reducing the manual inventory count. Typically, the initial cost for such robots will be high due to complexity of the system. However, the run time costs will be reduced as the labor required to do the inventory will be reduced. This technology is in research stages currently in India and companies like iFuture robotics are at the forefront to develop a cost-effective solution.

4. Amazon air prime: This is the advanced delivery system developed by AMAZON that is used to deliver the products to customers through the means of UAV's. The drones usually make their way through automated tracks and lift on to open sky and autonomously deliver to the customers. This was first implemented in 2016 in the UK and based on the success the company looks to expand the trial on a larger scale.

5. Robomart: It is a self- driving store that focuses on transferring of goods that are pre bought online. Robomart is designed to be completely autonomous but also has the additional feature of being completely electric.

Some of the current solutions that are incorporated in IISC's smart manufacturing CEFC facility are shown in the **Figures 3–5**.

Figure 3 shows the robot arm that is used to distinguish and sort things from the conveyor. The robot has an interactive human machine interface through which suitable instruction to the robot can be provided. The instructions provided are codes that can be suitably uploaded to the interface. Further the stabilizer provided to the robot arm will ensure the smooth operation of the robot. **Figures 4** and **5** shows the

Figure 3.
Yasakawa Robotic arm used for to pick and sort different items from a conveyor.

Figure 4.
Autonomous mobile vehicle/Hanhwa Cobot.

autonomous robot used for material handling process throughout the factory setup. There magnetic tapes that are provided along the floor of the smart factory which will guide the robot through the different parts of the factory. Further the robot is provided with obstacle detection system that will simply stop if the robot senses an obstacle along the path of motion. Further to ensure the safety of the human

Figure 5.
Autonomous mobile vehicle/Hanhwa Cobot system collaborating with robotic arm for material handling process.

operators there are emergency brakes that are provided to the system. The motion path for the robot can be suitably programmed according to the requirements and can be uploaded to the interactive human machine interface.

7. Conclusion

The growth of any country is dependent on the investments that the country makes in the manufacturing and the technological upgrade. This report deals with how the technology, the advent of Industry 4.0, and demands of the Indian market make India a viable choice. There are many viable entities like China, which act as a main competitor by the fact that it spends around 53% of the GDP in the manufacturing. There is a high level of foreign direct investment that the government of China has encouraged to improve the manufacturing and warehousing scenarios. However, there are certain limitations like the lack of good infrastructure for the manufacturing industry, IPR violations, etc. that the industries face making China not a viable option for companies to look at. In such a scenario, the Indian markets emerge as a suitable alternative where the service sector accounts to high amount of the overall GDP of the country at around 53% in [12]. Additionally, the affordability of the technology, the initiative taken by the Indian government makes it a suitable alternative for the companies to look at. Further concept of IOT is looked at as an important factor and is expected to capture close to 20% of the overall IOT share of the globe. Government of India looks at the initiatives of green corridor, setting up of isolated warehousing environments and make in India initiative encourages company to setup shops in India. Due to these initiatives India looks at leapfrogging the world's major players like China, Germany, USA as a major technological player for the industries to look at. Further for any industry to succeed the labor has to be skilled which India has in abundance. This report looks at the different concepts that may affect a typical warehousing scenario and how it affects the overall performance. Human machine interaction will play an important factor in determining the efficiency of the process. Further as a prospective solution the concept of least action is proposed where there is an optimal solution from the infinite paths. The proposed theory is simple and can be used for real time obstacle detection for a complex environment. The proposed concept can be looked at for a multi-robot system and IOT can be used for better control of the proposed robotic system in the warehouse.

Acknowledgements

We would like to thank, "Common Engineering Facility Centre (CEFC) IISc, Bangalore; Centre for Product Design and Manufacturing (CPDM), Department of Heavy Industries (DHI), Govt. of India," for their support in this endeavour.

Author details

Prajwal Prabhu, Subin Raj and Abhra Roy Chowdhury*
Division of Mechanical Sciences, Indian Institute of Science, Bangalore, India

*Address all correspondence to: abhra@iisc.ac.in

References

[1] Common Engineering Facility Centre (CEFC) IISc, Bangalore; Centre for Product Design and Manufacturing (CPDM), Department of Heavy Industries (DHI), Project Brochure report

[2] Brauner P, Runge S, Groten M, Schuh G, Ziefle M. Human factors in supply chain management. In: Yamamoto S, editors. Human Interface and the Management of Information. Information and Interaction for Learning, Culture, Collaboration and Business. HIMI. 2013. Lecture Notes in Computer Science. Vol. 8018. Berlin, Heidelberg: Springer; 2013

[3] Grosse EH, Glock CH, Neumann WP. Human Factors in Order Picking System Design: A Content Analysis. Elsevier; 2015. pp. 320-325

[4] Available from: https://content.knightfrank.com/research/677/documents/en/india-warehousing-and-logistics-india-warehousing-market-report-2018-5326.pdf

[5] Available from: https://www.roboticsbusinessreview.com/supply-chain/autonomous-mobile-robots-changing-logistics-landscape/

[6] Joo T, Shin D. Formalizing human–machine interactions for adaptive automation in smart manufacturing. IEEE Transactions on Human-Machine Systems. 2019;**49**(6):529-539

[7] Available from: https://www.therobotreport.com/autonomous-navigation-design-challenges/

[8] Available from: https://cds.cern.ch/record/101498/files/Thesis-1942-Feynman.pdf

[9] Available from: https://www.physics.mcgill.ca/~hilke/719.pdf

[10] Available from: https://www.dhl.com/content/dam/downloads/g0/about_us/logistics_insights/dhl_trendreport_robotics.pdf

[11] Available from: https://www.supplychaindive.com/news/4-types-of-autonomous-mobile-robots-and-their-warehouse-use-cases/529548/

[12] Available from: https://www.gita.org.in/Attachments/Reports/India%E2%80%99s%20Readiness%20for%20Industry%204.0.pdf

Section 2

Artificial Intelligence-of-Things (AIoTs)

Chapter 4

Applications of Machine Learning in Healthcare

Christopher Toh and James P. Brody

Abstract

Machine learning techniques in healthcare use the increasing amount of health data provided by the Internet of Things to improve patient outcomes. These techniques provide promising applications as well as significant challenges. The three main areas machine learning is applied to include medical imaging, natural language processing of medical documents, and genetic information. Many of these areas focus on diagnosis, detection, and prediction. A large infrastructure of medical devices currently generates data but a supporting infrastructure is oftentimes not in place to effectively utilize such data. The many different forms medical information exist in also creates some challenges in data formatting and can increase noise. We examine a brief history of machine learning, some basic knowledge regarding the techniques, and the current state of this technology in healthcare.

Keywords: machine learning, healthcare, big data, medicine, genetics, disease

1. Introduction

The advent of digital technologies in the healthcare field is characterized by continual challenges in both application and practicality. Unification of disparate health systems have been slow and the adoption of a fully integrated healthcare system in most parts of the world has not been accomplished. The inherent nature and complexity of human biology, as well as the variation between individual patients has consistently shown the importance of the human element in diagnosing and treating diseases. However, advances in digital technologies are no doubt becoming indispensable tools for healthcare professionals in providing the best care for patients.

The improvement of data technologies, including storage size, computational power, and data transfer speeds, has enabled the widespread adoption of machine learning in many fields—healthcare included. Due to the multivariate nature of providing quality healthcare to an individual, the recent trends in medicine have emphasized the need for a personalized medicine or "precision medicine" approach to healthcare. The goal of personalized medicine is to use large amounts of healthcare data to find, predict, and analyze diagnostic decisions, which physicians can in turn implement for each individual patient. Such data includes but is not limited to genetic or familial information, medical imaging data, drug combinations, population wide patient health outcomes, and natural language processing of existing medical documentation.

We will focus primarily on three of the largest applications of machine learning (ML) in the medical and biomedical fields. As a rapidly evolving field, there is a wide range of potential applications of machine learning in the healthcare field which may encompass auxiliary aspects of the field such as personnel management, insurance policies, regulatory affairs, and much more. As such, the topics covered in this chapter have been narrowed down to three common applications of machine learning.

The first is the use of machine learning in medical images such as magnetic resonance imaging (MRIs), computerized axial tomography (CAT) scans, ultrasound (US) imaging, and positron emission tomography (PET) scans. The result of these imaging modalities is a set or series of images which typically requires a radiologist to interpret and make a diagnosis. ML techniques have rapidly been advancing to predict and find images which may indicate a disease state or serious issue.

The second is natural language processing of medical documents. With the push towards electronic medical records (EMR) in many countries, the consensus from many healthcare professionals has been that the process is slow, tedious, and, in many cases, completely botched. This can sometimes lead to poorer overall healthcare for patients. One of the major challenges is the amount of physical medical records and documentation that already exists in many hospitals and clinics. Different formatting, hand-written notes, and a plethora of incomplete or non-centralized information has made the switch to adopting electronic medical records less than efficient.

The third machine learning application encompasses the use of human genetics to predict disease and find causes of disease. With the advent of next-generation sequencing (NGS) techniques and the explosion of genetic data including large databases of population-wide genetic information, the attempt to discern meaningful information of how genetics may affect human health is now at the forefront of many research endeavors. By understanding how complex diseases may manifest and how genetics may increase or decrease an individual person's risk can aid in preventative healthcare. This could provide physicians with more information on how to tailor a specific patients' care plan to reduce the risk of acquiring more complex diseases.

The common issue present in all three of these topics is how to translate health data acquired from the Internet of Things, into understandable, useful, trustworthy information for patients and clinician. How do we interpret hundreds of thousands of inputs and parameters from the data? How do we do this efficiently? What is the progress of addressing this problem currently?

2. Artificial intelligence and machine learning

Artificial intelligence (AI) has been intricately linked to the rise of modern-day computing machines. Machine learning has its roots and beginnings firmly planted in history. Alan Turing's work in cracking the German Enigma machine during World War II became the basis for much of modern computer science. The Turing Test, which aims to see if AI has become indistinguishable from human intelligence, is also named after him [1, 2].

At the height of the Second World War, the Allies had a significant logistical hurdle in the Atlantic. The United States and United Kingdom needed to set up secure shipping lines to move both armaments and troops to England in preparation for a mainland European invasion. However, the German U-boats were extremely effective at disrupting and sinking many of the ships traversing these shipping lanes [3]. As such, the Allies needed to intercept German communications to swing the

Battle of the Atlantic in their favor. The Germans encrypted their communications with The Enigma Machine, the most sophisticated encryption device of its time.

Turing and the rest of Bletchley Park were tasked with breaking the coded messages produced by The Enigma Machine and eventually produced The Bombe, a mechanical computing device which successfully decoded the cipher of The Enigma machine (**Figure 1**). Using the Bombe, they read the German orders sent to submarines and navigated their ships around these dangers. This was Turing's first intelligent machine. Alan Turing would later go on to describe the idea of a thinking machine which would eventually be called AI [4].

Machine learning is a subset of AI and the term was coined in the late 1950s by Arthur Samuel who published a paper on training computers to play checkers when he worked with IBM [5]. AI is best described as giving human-like intelligence to machines in a manner that directly mimics the decision making and processing of the human conscience. ML is the subset of AI that focuses on giving machines the ability to learn in an unaided manner without any human intervention.

By the late 1960s, researchers were already trying to teach computers to play basic games such as tic-tac-toe [6]. Eventually, the idea of neural networks, which were based on a theoretical model of human neuron connection and communication, was expanded into artificial neural networks (ANNs) [7, 8]. These foundational works laid dormant for many years due to the impracticality and poor performance of the systems created. Computing technology had not yet advanced enough to reduce the computational time to a practical level.

The modern computer era led to exponential increases in both computational power and data storage capacity. With the introduction of IBM's Deep Blue and Google's AlphaGo in recent decades, several leaps in AI have shown the capacity of

Figure 1.
Picture of the German Enigma machine which was used to code military communications. Taken from Wikimedia Commons.

AI to solve real world, complex problems [9, 10]. As such, the promise of machine learning has taken hold in almost every sector imaginable.

The widespread adoption of machine learning can be mostly attributed to the availability of extremely large datasets and the improvement of computational techniques, which reduce overfitting and improve the generalization of trained models. These two factors have been the driving force to the rapid popularization and adoption of machine learning in almost every field today. This coupled with the increasing prevalence of interconnected devices or the Internet of Things (IoT) has created a rich infrastructure upon which to build predictive and automated systems.

Machine learning is a primary method of understanding the massive influx of health data today. An infrastructure of systems to complement the increasing IoT infrastructure will undoubtedly rely heavily on these techniques. Many use cases have already show enormous promise. How do these techniques work and how do they give us insight into seemingly unconnected information?

2.1 Machine learning algorithms

Machine learning is broadly split into supervised and unsupervised learning. Algorithms falling under both categories implement mathematical models. Each algorithm aims to give computers the ability to learn how to perform certain tasks.

2.1.1 Supervised learning

Supervised learning typically employs training data known as labeled data. Training data has one or more inputs and has a "labeled" output. Models use these labeled results to assess themselves during training, with the goal of improving the prediction of new data (i.e., a set of test data) [11]. Typically, supervised learning models focus on classification and regression algorithms [12]. Classification problems are very common in medicine. In most clinical settings, diagnosing of a patient involves a doctor classifying the ailment given a certain set of symptoms. Regression problems tend to look at predicting numerical results like estimated length of stay in a hospital given a certain set of data like vital signs, medical history, and weight.

Common algorithms included in this supervised learning group are random forests (RF), decision trees (DT), Naïve Bayes models, linear and logistic regression, and support vector machines (SVM), though neural networks can also be trained through supervised learning [13]. Random forests are a form of decision trees but are an ensemble set of independently trained decision trees. The resulting predictions of the trees are typically averaged to get a better end result and prediction [14]. Each tree is built by using a random sample of the data with replacement and at each candidate split a random subset of features are also selected. This prevents each learner or tree from focusing too much on apparently predictive features of the training set which may not be predictive on new data. In other words, it increases generalization of the model. Random forests can have hundreds or even thousands of trees and work fairly well on noisy data [15]. The model created from aggregating results from multiple trees trained on the data will give a prediction that can be assessed using test data (**Figure 2**).

A method used to improve many supervised algorithms is known as gradient boosting. Taking decision trees as an example, the gradient boosting machine as it is commonly known, performs a similar ensemble training method as the random forest but with "weak learners." Instead of building the decision trees in parallel as in the random forest algorithm, the trees are built sequentially with the error of the previous tree being used to improve the next tree [16]. These trees are not nearly as deep as the random forest trees, which is why they are called "weak" (**Figure 3**).

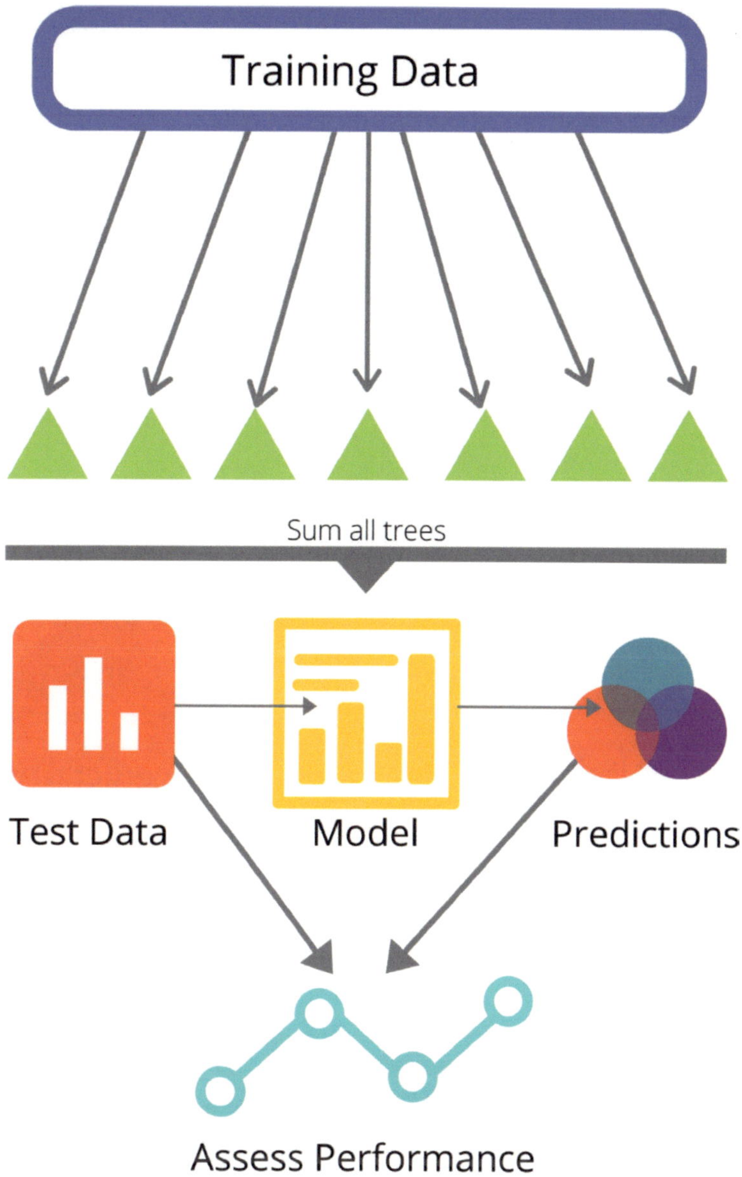

Figure 2.
Example of a workflow for training and assessing a random forest model. Each green triangle represents an independently trained tree from the training data. The prediction of each tree is summed and is represented as the model. Test data is then fed to the model, i.e., all the trees, and the resulting prediction is made. The prediction is then compared to the original test data to assess how the model performs.

Typically, better results can be achieved with gradient boosting, but tuning is much more difficult, and the risk of overfitting is higher. Gradient boosting works well with unbalanced data and training time is significantly faster due to the gradient descent nature of the algorithm [17, 18].

2.1.2 Unsupervised learning

Unsupervised machine learning uses unlabeled data to find patterns within the data itself [19]. These algorithms typically excel at clustering data into relevant groups, allowing for detection of latent characteristics which may not be

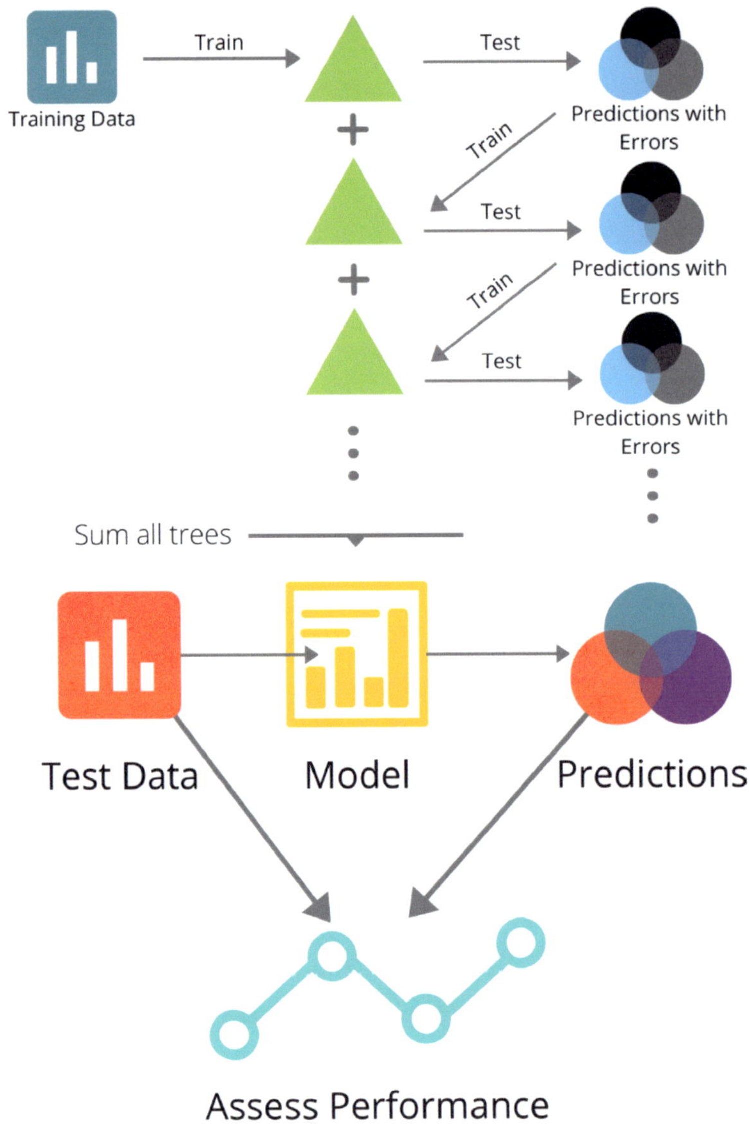

Figure 3.
Example of a simple workflow for training and assessing a gradient boosting machine model. Each green triangle represents a trained tree from the training data with the subsequent tree using the residuals or errors from the prior tree to improve its prediction. The prediction of each tree is summed and is represented as the model. Test data is then fed to the model, i.e., all the trees, and the resulting prediction is made. The prediction is then compared to the original test data to assess how the model performs.

immediately obvious. However, they are also more computationally intensive and require a larger amount of data to perform.

The most common and well-known algorithms are K-means clustering and deep learning, though deep learning can be used in a supervised manner [12, 20]. Such algorithms also perform association tasks which are similar to clustering. These algorithms are considered unsupervised because there is no human input as to what set of attributes the clusters will be centered on.

The typical k-means algorithm has several variations such as k-medians and k-medoids, however the principle is the same for each algorithm. The algorithm uses Euclidian distance to find the "nearest" center or mean for a cluster assuming there are *k* clusters. It then assigns the current data point to that cluster and then recalculates the center for the cluster, updating it for the next data point [21]. The biggest drawback to this algorithm is that it must be initialized with an expected number of "means" or "centers." Improper selection of the *k* value can result in poor clustering.

Deep learning uses neural nets to perform predictions even on unlabeled data as well as classification techniques. Based off models of human neurons, perceptrons, as they are typically called, are organized into many networked layers making the network "deep" in nature [20]. Each perceptron has multiple inputs and a single output. They are organized into layers where the outputs of the previous layer serve as the inputs for the next layer. The input layer requires one perceptron per input variable and the subsequent layers are determined before training by a human (**Figure 4**). This is one of the difficulties and challenges in building an effective neural net. The computationally intensive nature of computing each perceptron for a large neural net can mean that training alone can take days to weeks for large data sets [22].

2.1.3 Hyperparameters

In machine learning, a model typically has a set of parameters as well as a set of hyperparameters. Parameters are variables about the model that can be changed during training. For example, parameters can be the values of the training data itself with each piece of data being different along one or several of the parameters. Whereas hyperparameters are typically set before training occurs and cannot change once learning begins. Hyperparameters typically are set to tune values like the model's learning speed and constrain the algorithm itself.

Different algorithms will have different sets of hyperparameters. For example, a common hyper parameter for artificial neural networks is the number of hidden layers. Additionally, a separate but related hyperparameter is the number of perceptrons in each hidden layer. Whereas a similar equivalent in decision trees would

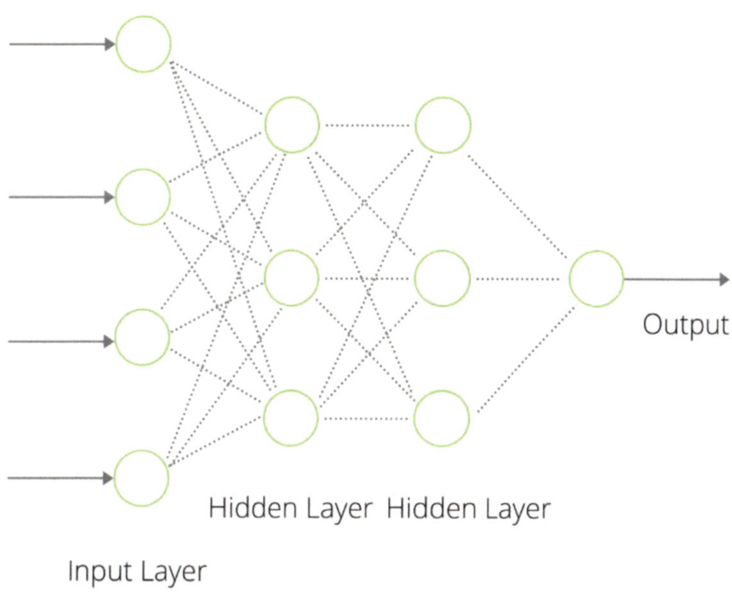

Figure 4.
Example of a simple neural net with two hidden layers of three perceptrons each. The number of inputs, number of hidden layers, and number of perceptrons in each layer can be changed. Additionally, the connections between layers and perceptrons can also be changed.

be the maximum number of leaves in a tree or the maximum depth for a tree. Other common hyperparameters include learning rate, batch size, dropout criterion, and stopping metric.

Properly selecting hyperparameters can significantly speed up the search for a proper generalized model without sacrificing performance. However, in many cases finding the proper set is more of an art than a science. Many researchers have attempted to make hyperparameter searching a more efficient and reproducible task [23–25]. Again, this process also highly depends on the algorithm, dataset, and problem you are trying to solve. A machine learning model can be tuned a nearly infinite amount of different ways to achieve better performance. Hyperparameters represent a way to reproduce results and also serve as a tool to properly validate models.

2.1.4 Algorithm principles

Considering the pace of research in the field, there are constant advances and improvements to many of these machine learning techniques, but the important thing to remember is that not all algorithms work for all use cases. Each algorithm has advantages and disadvantages. Certain data types may also affect the performance of individual algorithms and the time spent implementing such models will often be a result of testing different variations, parameters, and hyperparameters within these algorithms to achieve the best generalized performance.

2.2 Assessment of model performance

The goal of any machine learning algorithm is to utilize real data to create a model that performs the best on real-world scenarios, and that can be assessed in a quantitative, reproducible manner. Assessment of statistical models is a whole subfield in itself, but we will briefly discuss the basics, which are applicable for almost any machine learning algorithm you will come across.

2.2.1 Sensitivity vs. specificity

Sensitivity and specificity are two important metrics used in a statistical or machine learning model to assess if the model is performing successfully. As such, it is important to understand what each of these numbers tell us about what a trained model can do, and what the model cannot do.

Sensitivity is the probability that a positive result occurs given that the sample is indeed positive. Mathematically,

$$Sensitivity = \frac{(Number\ of\ True\ Positives)}{Number\ of\ True\ Positives + Number\ of\ False\ Negatives}$$

This is also sometimes referred to as the recall or hit rate, or just simply the true positive rate, and the sensitivity is equivalent to 1 - *False Negative Rate*.

Specificity is the probability of a negative result given that the sample is negative. Mathematically,

$$Specificity = \frac{Number\ of\ True\ Negatives}{Number\ of\ True\ Negatives + Number\ of\ False\ Positives}$$

This value is also referred to as the selectivity of the test. This is equivalent to 1 - *False Positive Rate*.

2.2.2 The receiver operator curve and area-under the curve

The standard metric for assessing the performance of machine learning models is known as the receiver operating characteristic (ROC). The ROC can be summarized by a number from 0 to 1, which is the measured area-under-the-ROC curve (AUC). The ROC curve is a 2D plot that measures the false positive rate vs. true positive rate. There are four numbers that are used to determine the effectiveness of a test: true positive rate, false positive rate, true negative rate, and false negative rate.

True positive and true negative are the correct answers to a test while false positive and false negative are incorrect answers to the test or model. These numbers can be condensed further into two numbers known as sensitivity and specificity. We have already discussed sensitivity and specificity but now we will discuss how they are used to create the ROC.

Ideally a test would have both high sensitivity and high specificity. However, there is a tradeoff, prioritizing one often leads to the detriment of the other. When setting the threshold low, one will receive a high true positive rate (high sensitivity) and a high false positive rate (low specificity). Conversely, setting the threshold high will result in a low true positive rate (low sensitivity) and a low false positive rate (high specificity).

The ROC and AUC metric is used to characterize most of the classification tasks many machine learning models are attempting to do; does this person have the disease or do they not? If a test has a high sensitivity and a high specificity it is considered a near perfect test and the AUC is close to 1 (**Figure 5**). If the test is random then the AUC is 0.5. The x-axis is typically the false positive rate (or 1 – specificity). Ideally, the false positive rate is as low as possible. The y-axis is typically the true positive rate (sensitivity). The sensitivity is what is usually maximized. On a typical curve, the midpoint of the curve is the most balanced trade-off between sensitivity and specificity though this is not always the case. The AUC value is a simpler, more generalized way, to assess the performance rather than the varying tradeoffs between sensitivity and specificity.

Another way to think of AUC is as a percentage the model can correctly identify and separate a positive result from a negative result. Given an unknown case, a model with an AUC of 0.75 has a 75% chance of correctly identifying whether the case is a positive case or a negative case. This number will quickly tell you the results of any model.

2.2.3 Overfitting

Overfitting is one of the main concerns when training any model [26]. Simply put, when training a model on a set of data, over-training the model will improve the performance of the model on that specific dataset but at the cost of losing generalization to other datasets. An overfitted model will not work when applied to new data it has never seen before. From a practical standpoint, such a model is not very useful in a real-world application.

When training any machine learning model, the ideal result is a *generalized* model. A generalized model works well on a variety of different cases and a variety of different datasets, especially data it has never seen before. As such, many researchers are hesitant to give too much credence to a model or method that utilizes a single dataset.

A variety of methods have been used to prevent models from overfitting and many of these are now encapsulated in the *hyperparameters* discussed earlier.

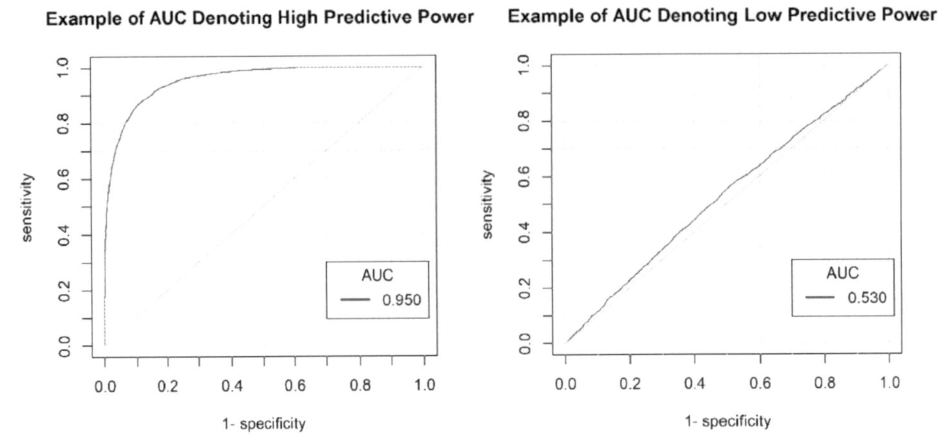

Figure 5.
Examples of an AUC denoting a model which has good predictive power (left) and an AUC denoting a model with poor or near random predictive power (right). 1 – Specificity is sometimes written as false positive rate (fpr) and sensitivity can be read as true positive rate (tpr).

The idea is to prevent the models from adapting too quickly to the dataset it is being trained on. This subset of methods is known as *regularization* [27].

One such method, used in neural nets, is called dropout. This method is widely used to prevent artificial neural nets from overfitting during classification tasks. The method is fairly simple. During the training process, random perceptrons and their corresponding connections are "dropped" from the network. These "thinned" networks have better performance compared to other regularization techniques on supervised learning tasks [28].

Often a method known as cross-validation is used to assess the performance and validate the *generalized* predictive ability of a model. The most common method for building machine learning models is the partitioning of the data set into roughly 80% for training and 20% for testing. This partition is typically less useful for linear models but splitting is more beneficial for complex models [29]. During cross-validation, this split is done in separate sections of the data to ensure proper coverage. For example, if a 10-fold cross-validation is performed, the first split in a data set with 100 observations could be a the first 80 for training and the last 20 for test, the second split could be the first 10 and last 10 for test and the middle 80 for training, etc. (**Figure 6**). This creates 10 models using the same algorithm just trained and tested on different portions of the same data. The average performance of these 10 models gives a good measurement of the generalized performance of the algorithm on that type of data.

2.3 Big data and the health information explosion

The healthcare sector has always had a very large amount of information, often times stored as physical documents in clinics, hospitals, regulatory agencies, and biomedical companies [30, 31]. With the push to electronic medical records (EMR), this information is rapidly being transformed into a form which can be leveraged by AI technologies. The estimated amount of healthcare data stored in 2011 was around 150 exabytes (1 EB = 10^{18} bytes), though that number is most likely exponentially larger almost a decade later [32, 33]. These large databases, when in a digitized form, are often known as Big Data.

However, such healthcare information is very different in both form and function. Visual data in the form of medical images is very different than familial history

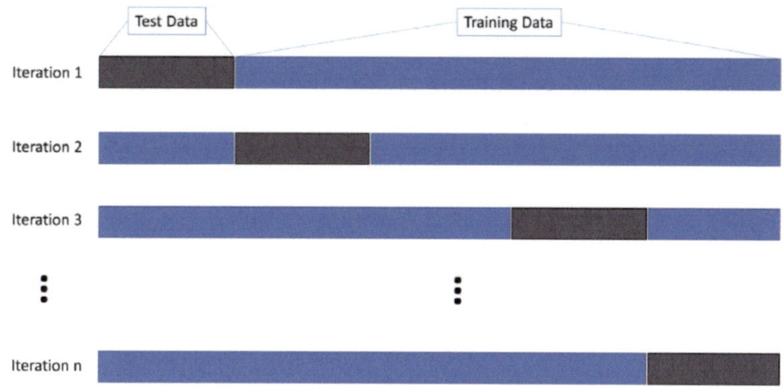

Figure 6.
Example of a set of cross validation splits. There are n splits for the number of iterations desired and the results of all iterations are averaged to assess the generalized performance of a model trained on a dataset.

which may be simple text-based information. Laboratory and clinical tests may be reported as numbers only, while health outcomes are often qualitative in nature and may be a simple yes or no entry in a spreadsheet. Insurance and administrative data is also indirectly linked to various information, such as patient outcomes, while information from sensor based technologies like EKGs, pulse oximeters, and EEG provide time-series data of vital signs [34].

Additionally, the genomic revolution has contributed enormously to the data explosion. Large-scale genetic databases such as the Cancer Genome Atlas (TCGA) and the UK Biobank include thousands of patients' genetic sequencing information along with various other health information such as disease state, age of diagnosis, time of death, and much more [35–38]. Copy number variation (CNV) data from the UK Biobank's roughly 500,000 patients, which does not even contain the raw sequence reads, is almost 2 Terabytes (TB) alone in flat text files. These genetic databases rely on an array of assays and sequencers spread across different hospitals

Database	Size of data	Number of participants	Status	Start date
The Cancer Genome Atlas	2.5 petabytes	11,300 [36, 41]	Completed	2005
The UK Biobank	~26 terabytes*,+	~488,377 [42]	Ongoing	2006
The European Prospective Investigation into Cancer and Nutrition (EPIC)	Unclear*	~521,000 [43]	Ongoing	1992
Estonian Genome Project	Unclear*	~52,000 [44]	Ongoing	2007
deCODE	Unclear*	~160,000 [45, 46]	Ongoing	1998
China Kadoorie Biobank	Unclear*	~510,000 [47, 48]	Ongoing	2004
Lifelines Cohort Study	Unclear*	~167,000 [49]	Ongoing	2006
All of Us (Precision Medicine Initiative)	Unclear*	Currently ~10,000, planned 1,000,000 [12, 50–52]	Ongoing	2015
FinnGen	Unclear*	Planned 500,000 [53]	Ongoing	2017

**Project is continuing to collect more data.*
+Number represents genetic data only. Project or study may also include unreported data including medical images and health records.

Table 1.
Overview of largest biobank databases as of 2019.

and research facilities around the globe, before being processed and transferred to their respective centralized storage databases [35, 39, 40].

The collection of biological data and creation of these databases show no evidence of slowing down. Many biobanks, databases which contain some form of biological samples such as blood or serum, contain thousands of participants and many have plans to collect hundreds of thousands of samples from patients (**Table 1**). Because many databases are growing so quickly it is unclear how much data resides in many of these databases. However, The Cancer Genome Atlas alone contains 2.5 petabytes (1 PB = 10^{15} bytes) of data and the UK Biobank contains 26 terabytes (1 TB = 10^{12} bytes) of just genetic information (UK Biobank also contains medical images such as brain scans which is not included in this table).

Implementing machine learning systems into a hospital with this complex information Web is usually slow, due to the abundance of caution needed to ensure patient health. Many physicians are also wary of adopting new systems that are unproven in a clinical setting due to the risk of litigation and potentially catastrophic consequences for their patients.

3. Machine learning of medical images

Modern medical images are digital in nature. To effectively utilize them in healthcare there are several challenges that must be overcome. Medical imaging describes a collection of techniques to create visual representations of interior portions of the human body for the purpose of diagnosis, analysis, and medical intervention. This is beneficial in avoiding or reducing the need for the older clinical standard of exploratory surgery. Since opening any portion of the human body through surgical means increasing the risk of infections, strokes, and other complications, medical imaging is now the preferred tool for initial diagnosis in the clinical setting.

The current clinical standard of assessing medical images is the use of trained physicians, pathologists, or radiologists who examine the images and determine the root cause of clinical ailments. This clinical standard is prone to human error and is also costly and expensive, often requiring years or decades of experience to achieve a level of understanding which can consistently assess these images. Considering that the demonstration of viable machine learning capabilities in the modern age was demonstrated by Andrew Ng using images pulled from YouTube videos, it is clear why medical images were one of the first areas addressed during the initial adoption of machine learning techniques in healthcare [54].

Accuracy of diagnosis is extremely important in the medical field as improper diagnosis could lead to severe consequences and results. If a surgery is performed where none was needed or a misdiagnosis leads to improper dosages of prescribed medication, the possibility of a fatal outcome increases. In the realm of image processing, most techniques rely fundamentally on deep learning (DL) and specifically in artificial neural networks (ANNs). Modern techniques utilize improvements to ANNs in the form of convolutional neural networks (CNNs) to boost performance when classifying images.

The majority of the current publications are using some form of CNNs when it comes to object detection in medical images [55]. Graphic-processing unit (GPU) acceleration has made the building of deep CNNs more efficient, however significant challenges in creating a competent model still exist. The biggest issue is the need for a large amount of annotated medical image data. The cost to aggregate and create such databases is often prohibitive since it requires trained physicians' time to annotate the images. Additionally, concerns involving patient privacy often hinders the ability to make such databases open-source. Many studies only use around 100–1000 samples in training CNNs. This limited sample size increases the risk of overfitting and reduces the accuracy of the predictions [56].

Concerns regarding the implementation of machine learning into clinical diagnosis have been raised regarding proper validation of models [57]. The main fears entail properly scoping the intended goals of a machine learning model, reducing dimensionality of the data, and reproducibility of training such models on real-world and new clinical data. Validating results on other datasets can be difficult due to the lack of larger datasets for niche diseases, where the aggregation of this data can take more work than the actual training of the model. Medical imaging data is inherently more difficult to acquire and is more difficult to store and process. The infrastructure to handle the data has simply not kept up with the increase in the amount of data.

3.1 Lesion detection and computer automated detection

The most common use of current machine learning technologies in medicine is for computer automated detection (CAD) specifically in the detection of lesions such as those commonly found in mammograms, brain scans, and other body scans [58]. These methods use CNNs to arrive at the probability that a candidate lesion is in fact a lesion, often utilizing several 2D slices of 3D rotational scans of either CAT or MRI images.

Ultrasound images are also used in training and a variety of methods such as randomized rotation of the images or centering candidate lesions in the center of the image. Especially in mammography, CAD techniques have reached a level where they are used as a "second opinion" for most radiologists, greatly improving the accuracy of screenings without doubling the cost associated with using a human as the "second opinion" **Figure 7**.

CAD is also currently split into detection and diagnosis. This distinction is subtle but important. A lesion can be categorized as either benign or malignant, based off a physician's knowledge and assessment. However, the actual detection is a crucial first step in treating a patient.

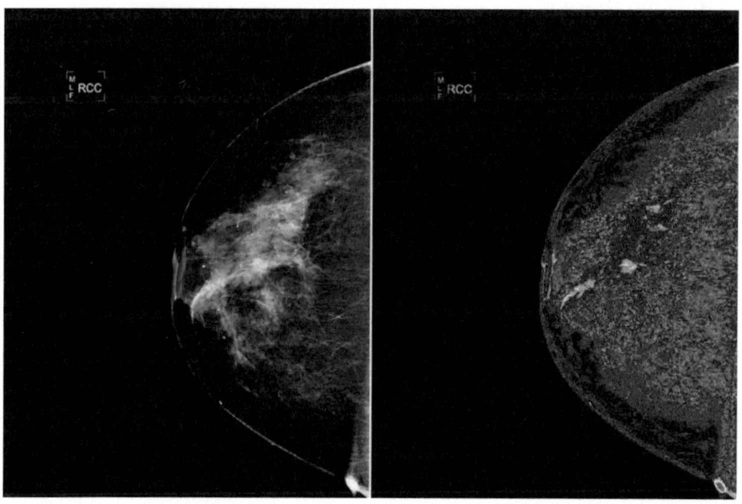

Figure 7.
Example of mammogram with the left image being that of a raw mammogram and the right hand being the image with the detection overlaid with the region of interest in white, using NASA software originally used to enhance earth science imagery. Taken from NASA press release, credited to Bartron Medical Imaging.

Computer aided detection is the actual recognition of potential lesions from a medical image. For example, detection and segmentation of glioblastoma is a difficult task, due to the invasive and widespread nature of these tumors. Unlike other brain tumors, they are not easily localized and assessing how treatments such as chemotherapy are performing is in itself a difficult task. Deep learning has aided in this by helping automate assessment of glioblastoma MRIs [59].

Computer aided diagnosis describes the probability a lesion is malignant in nature. These methods are primarily used to improve the accuracy of diagnosis and improve early diagnosis in the clinical setting. Again, these tasks have consistently been performed by machine learning especially in brain related applications, due to the difficult nature of assessing brain health. Additionally, diagnosis of Alzheimer's through medical imaging is a possible application for deep learning which is showing some promise [60, 61].

4. Natural language processing of medical documents and literature

Electronic medical records (EMR), the new standard in many hospitals, require complex digital infrastructure. Unification of health data in a formatted manner is a major goal as it should increase the efficiency of hospitals as well as improve patient health outcomes. However, a significant problem is the historical existing physical documentation. Transferring these existing documents into an electronic form is difficult and would be very tedious and expensive if people were hired to manually input such information into an electronic system.

One application of machine learning, which may aid in this problem, is natural language processing (NLP). By scanning these documents rapidly and integrating the resulting images into a database, these systems attempt to extract readable data from free text and incorporates image processing to identify key words and terms. Handwritten physician notes contain information such as patient complaints, the physicians own observations, and patient family history. This clinical information can be annotated. However, poorly worded or inaccurate writing by the physician can make it difficult to accurately assign this information to appropriate categories. Forms and documents that already have structure make for much easier language processing, though there is still the risk of missing data **Figure 8**.

Creating a system for improved clinical decision support (CDS) with old patient records is feasible. Any such system is structured to aid in clinical decision making for individual patients based on a database of computerized knowledge. Such a system could be envisioned as two-fold: 1. extracting facts about the patient from their medical record, either through written or typed physician notes or labs or dictation involving audio NLP, 2. Associating possible disease states based on extracted information from previous known cases or through literature search via NLP [62]. Integration of several specialized NLP systems is required for any true and practical implementation of such a CDS system.

Likewise, compilation of the existing scientific research into central repositories is a difficult task. Sometimes physicians may be unaware of a promising new treatment just due to the difficulty of parsing the tidal wave of new papers. Scientific publications have always been widely dispersed across multiple journals and the modern-day information explosion has only exacerbated the issue. When it comes to compiling information such as results from genome-wide association studies (GWAS), the primary method has been a manual curation of the information by certain individuals within the scientific community: "librarians" so to speak.

Recently, a paper published in Nature Communications used machine learning systems to automatically compile GWAS information from open-access publications

Nursing Care Plan

CLIENT ID:
NAME:
D.O.B.:
DOCTOR:
PENSION:

LIFESTYLE ISSUES	GOAL OF CARE	CARE OR INTERVENTION REQUIRED Tick and/or Highlight Appropriate Response
Links to Assessment: Baseline Health Assessment *Communication assess (11-04)*		Hearing Hearing loss: Partial ☐ Profound ☐ Right ☐ Left ☐ Both ☐ Aids used: Behind the ear ☐ Inside the ear ☐ Right ☐ Left ☐ Both ☐ Storage details: .... Assistance Required? ☐ Yes ☐No Are hearing aids worn ☐ Yes ☐No Battery type ☐..........
Notes		

LIFESTYLE ISSUES	GOAL OF CARE	CARE OR INTERVENTION REQUIRED Tick and/or Highlight Appropriate Response
SKIN INTEGRITY: **Links to Assessment:** *Waterlow Risk Assessment (11-06)* *Wound Management-Complex (11-07a)* *Wound Management-Simple (11-07b)*	**RESIDETS SKIN INTEGRITY IS MAINTAINED AT OPTIMUM LEVEL**	☐ Bed/Chair bound:.......... ☐Repositioning frequency ☐2nd hourly ☐3rd hourly ☐4th hourly ☐Other ☐ Aid (in Bed):.......... ☐ Aid (on chair):.......... ☐ Creams:.......... ☐ Leg Protectors:.......... ☐ Bed Rail Protectors: Yes☐ No☐ ☐ Other:.......... **Refer to Wound Management Charts as necessary**

Name	**Designation**
Signature	**Date**
Notes	

Figure 8.
Example of a nursing care plan which represents a formatted health document. Most of these plans were filled out by hand and many hospitals have transitioned such forms to electronic records. However, older documents still need to be transferred to digital form. Taken from Wikipedia commons.

and extract GWAS associations into a database with the aim of helping curators. Though the results are somewhat inconsistent (60–80% recall and 78–94% precision) it represents one of the many ways NLP is being utilized to aid in medical discovery [63].

4.1 Examples of natural language processing in healthcare research

There are many exciting possibilities where NLP could be used to improve medicine and medical research. We will discuss a few interesting findings with similar approaches but different goals. This is by no means an expansive list but highlights the broad spectrum of possible machine learning applications.

In 2015, a research group published a paper reporting 100% accuracy of predicting onset of psychosis using recorded dialog of clinically high-risk youth. Each youth was interviewed over a period of 2.5 years every 3 months. Based on the transcripts of these interviews, a machine learning algorithm was trained to predict whether a patient would develop psychosis. This was done using what is known as Latent Semantic Analysis to determine coherence of speech using NLP. The sample size for this study was rather small however (n = 34) [64].

Another study used NLP to identify cirrhosis patients and risk-stratify the patients. This study was able to correctly identify cirrhosis patients from electronic health records, ICD-9 code combinations, and radiological scans with a 95.71% sensitivity and 93.88% specificity [65]. This indicates that such a system could correctly identify cirrhosis patients based off existing medical data in most hospitals.

Yet another study used NLP to accurately identify reportable cancer cases for national cancer registries. This method analyzed pathology reports and diagnosis codes to identify patients with cancer patients using supervised machine learning. The accuracy was 0.872 with a precision of 0.843 and sensitivity of 0.848 [66]. The primary goal of this study was to automate the process of reporting cancer patients to the National Program of Cancer Registries in the United States.

These examples of NLP use in healthcare highlight the wide diversity of applications within medicine. Language is the primary means of communicating complex information, doctors' notes and annotated medical documents hold valuable insights in populations and individual patient health. The irregularity and variance of language and extraction of higher-level information into relevant subcategories makes analysis difficult. Machine learning is showing promising results in performing such complex analyses.

5. Machine learning in genetics for the prediction and understanding of complex diseases

Genetic information and technologies have exploded since 2008, creating difficult challenges in how to handle the exponentially increasing data. Advances in genetic sequencing speed, namely NGS technologies have exponentially increased the speed at which a whole human genome is sequenced, while also dramatically reducing costs. The human genome is a complex physical structure that encodes all the information of human development and characteristics. The genome is highly interconnected and deciphering most of these instructions is still a mystery to us. Variation of genomes between people also increases the complexity of understanding gene interactions.

Many health initiatives have focused on acquiring large sample sizes of human genomes to help identify statistically relevant trends among different populations of humans. However, the 23 chromosomes of the human genome contain around 20,000 genes which have been identified as the primary coding sequences for the proteins necessary in building the biological components of our cells [67]. This number is still a rough estimate and some estimates indicate that there may be as many as 25,000 genes or as few as 19,000 [68, 69]. A large swathe of genetic information that does not code for any proteins is not included in these estimates.

A growing body of literature indicates that certain sections of what has been colloquially called *genetic dark matter*, or *missing heritability*, exists [70–74]. These terms refer to the portions of DNA which have no apparent protein coding function, but may be relevant to the level of gene expression in a person's genetic code [75, 76]. Levels of gene expression may cause protein overload or deficiency, which can lead to a variety of health problems. Additionally, structural differences in the physical structure of how the DNA is bound into chromosomes and then subsequently unwrapped during both the duplication process and translation and transcription process, can also affect the level of gene expression.

For example, methylation or acetylation of the DNA backbone can make it more difficult (methylation) or easier (acetylation) to unravel the DNA strand during normal cell processes like replication or protein assembly. Evidence of multiple copies of the same gene have also been classified in what is described as copy number variations (CNV) which indicate duplication, triplication, and deletion events of certain areas of the genome in an individual. Understanding this highly interconnected and nonlinear relationship between all the different of the areas of the human genome is difficult.

With machine learning, scientists have begun to find patterns and trends which can be modeled in a more predictable manner. Utilizing the ever-growing amount of genetic data, machine learning has the potential of accurately predicting who is at risk of acquiring certain diseases such as cancers and Alzheimer's disease. Mental illnesses such as schizophrenia and bipolar disorder have also been known to run in families, indicating a possible genetic link.

5.1 Inherited vs. environmental risk

Disease risk can be broadly categorized into inherited risk and environmental risk. Inherited risk describes a person's disposition to acquiring complex diseases due to a trait which is genetically passed down from their predecessors. This includes genetic mutations contained within their germline DNA which may predispose them to cancers or other health conditions [77, 78].

Environmental risk describes somatic mutations, or mutations to a person's DNA due to something they have encountered in their environment. These mutations can still increase a person's risk of acquiring a disease but they do not affect the germline, and will not be passed on to their progeny and thus will not be inherited [79].

Inherited risk describes mutations that exist in the human germline and which will be passed onto the offspring through normal reproduction. Whereas, somatic mutations may affect organs or a set of cells, germline mutations exist in all the cells of the offspring. Many of these mutations may be passed through paternal lineage and there is some indication that certain individuals may have disease predisposition but which cannot be directly linked to familial history but could still be due to these hidden germline mutations [80–82].

Several different types of mutations may exist within a human genome. They are broadly categorized as single nucleotide polymorphisms (SNPs), structural variations or copy-number variations (CNVs), and epigenetic variations.

SNPs are a single or point mutation of one base pair in the human genome that occurs in at least 1% of the human population [83, 84]. These mutations are the most common source of genetic variation and can occur both within coding regions and outside of coding regions of the genome. SNPs contribute to vast differences even between relatives and can arise because of both inheritance and development in the womb. Within SNPs there are common and rare variants, with rare variants occurring less than 0.5% within the global sample [84].

Structural variations and specifically CNVs are deletions, insertions, duplications, and inversions of large regions of DNA. These structural differences are usually inherited and a typical human can have anywhere between 2100 and 2500 structural variations [84]. These variations were found to cover more of the human genome than SNPs alone [83].

Epigenetic variation describes variations in the chemical tags attached to DNA or associated structures such as histones, which affects how genes are read and activated. Epigenetics includes DNA methylation and acetylation, histone modifications, and non-coding RNAs which all affect the degree to which a gene may be expressed [85]. As a newer field, it is unclear how much of these epigenetic

Type of cancer	Cases	Controls	AUC
Breast invasive carcinoma (men and women)	977	8821	0.81
Glioblastoma multiforme	484	9314	0.86
Ovarian serous cystadenocarcinoma	424	4268	0.89
Thymoma	111	9687	0.78
Uveal melanoma	80	9718	0.80

Table 2.
Sampling of performance of GBM models trained on data from the Cancer Genome Atlas.

variations are inherited from generation to generation, and how much is a result of environmental factors [86].

5.2 Prediction of cancers through germline copy number variations

One of the exciting methods we have discovered is the utilization of germline copy number variations in the prediction of different cancers. We have found that it is possible to use machine learning models, specifically gradient boosting machines (GBM), a form of decision trees (DT), to predict whether a person has a particular cancer. The models created were able to predict cancers such as ovarian cancer (OV) and glioblastoma multiforme with an AUC of 0.89 and 0.86 respectively [87], using copy number variation data taken from germline blood samples only. This result indicates that there is a significant inherited portion contributing to cancer risk in many, if not all cancers. Since these CNVs are also taken from germline DNA, the likelihood of continued inheritance to future generations is high.

This method does not look solely at SNPs as many previous methods rely on [88]. Most SNP data specifically looks at mutations within protein coding genes while ignoring the rest of the genome, whereas our method utilizes a whole genome approach by averaging the copy numbers of a person's entire genome as the basis for predicting cancer. Copy number variation accounts for a large amount of human genetic diversity and is functionally significant though the exact mechanisms are still unclear [77, 83].

These results demonstrate that almost all cancers have a component of predictability in germline CNVs which can be used to predict an individual's risk to acquiring that cancer **Table 2**. Experiments were performed on two independent databases: The Cancer Genome Atlas and the UK Biobank. The first database contains about 10,000 individuals and latter contains about 500,000 individuals.

Future studies may improve on the performance and the models could potentially be used as a tool to assess individual risk for diseases. Since the method can also be easily generalized to other diseases, we anticipate work to continue to encompass other potentially complex diseases which may have inherited components to them.

6. Conclusions

Application of digital technologies such as machine learning in the healthcare field is entering an exciting era. The collision of informatics, biology, engineering, chemistry, and computer science will rapidly accelerate our knowledge of both hereditary and environmental factors contributing to the onset of complex diseases. The potential of utilizing copy number variations in the prediction of cancer diagnosis is exciting. Utilizing machine learning to create an interpretable method of understanding how the genomic landscape interlinks across genes to contribute to inherited cancer risk could potentially improve patient healthcare on an individual level.

Databases such as The Cancer Genome Atlas and UK Biobank are invaluable resources, providing high statistical power to scientific analysis. As other large-scale population data projects near completion in the coming decade, the methods laid on the foundation of The Cancer Genome Atlas and UK Biobank will continue to benefit and improve as sample sizes easily begin to move into the regime of millions of patients. Tracking populations around the world will truly aid in the goal of precision medicine.

Natural language processing will be essential in improving the practicality of translating scientific findings and results of other machine learning methods into a clinical setting. Multiple specialized systems will have to be integrated with each other to effectively extract the wealth of information into a format which can be utilized effectively by physicians and healthcare professionals.

Image analysis is becoming a staple in many diagnostic endeavors and will continue to improve the accuracy of radiological diagnosis. Detection of malignant masses and validation and verification of existing diagnosis has the potential to improve patient outcomes, while reducing errors. As a non-invasive method of looking inside the human body, any improvements in healthcare imaging will reduce the need for risky or ill-informed operations that could lead to other complications such as infections and blood clots.

The examples discussed in this chapter are some of the most promising works in applying machine learning in the healthcare field. Resolving big health data into a usable form will undoubtedly require machine learning techniques to improve. Infrastructure to support such learning techniques is currently not stable or standardized. Bringing such methods from concept to practical clinical use is contingent on both validation of these results and an appropriate infrastructure to support it.

A large variety of devices and storage methods will need to be unified and standardized to benefit from the increased data collection. Information about how human genetic variation can contribute to individual susceptibility allows patients and doctors to make early lifestyle changes in a preventative manner. Likewise, it can inform physicians of which types of prognostics and diagnostics would be the most relevant for a specific patient, saving both time and money, while improving patient outcomes in the long term. Just as AI started with Turing decoding the enigma machine, we are now going to use AI and machine learning to decode the secrets of the human body and genome.

Conflict of interest

The corresponding author is a distant relative of the editor of this book.

Notes/thanks/other declarations

The author would like to thank the University of California, Irvine for support during the writing of this chapter.

Author details

Christopher Toh* and James P. Brody
University of California, Irvine, United States of America

*Address all correspondence to: tohc@uci.edu

References

[1] Copeland J. The turing test. Minds and Machines. 2000;**10**(4):519-539

[2] French RM. The turing test: The first 50 years. Trends in Cognitive Sciences. 2000;**4**(3):115-122

[3] Edwards S. World war II at sea: A global history. The Journal of American History. 2019;**106**(1):237

[4] Turing AM. Computing machinery and intelligence. Parsing the Turing Test: Philosophical and Methodological Issues in the Quest for the Thinking Computer. 2009. p. 23-65

[5] Samuel AL. Some studies in machine learning. IBM Journal of Research and Development. 1959;**3**(3):210-229

[6] Samuel AL. Programming computers to play games. Advances in Computers. 1960;**1**(C):165-192

[7] Fukushima K. Cognitron: A self-organizing multilayered neural network. Biological Cybernetics. 1975;**20**(3-4):121-136

[8] Fukushima K. Neocognitron: A self-organizing neural network model for a mechanism of pattern recognition unaffected by shift in position. Biological Cybernetics. 1980;**36**(4):193-202

[9] Huang K, Mashl RJ, Wu Y, Ritter DI, Wang J, Oh C, et al. Pathogenic germline variants in 10,389 adult cancers. Cell. 2018

[10] Silver D, Hassabis D. AlphaGo: Mastering the ancient game of go with machine learning. Google Research Blog. 2016

[11] Brewka G. Artificial intelligence—a modern approach by Stuart Russell and Peter Norvig, Prentice Hall. Series in Artificial Intelligence, Englewood Cliffs, NJ. The Knowledge Engineering Review. 1996;**11**(1): 78-79

[12] Alpaydin E. Introduction to Machine Learning. London: The MIT Press. 2014;**3**:640

[13] Kotsiantis SB. Supervised machine learning: A review of classification techniques. Informatica (Ljubljana). 2007;**31**(3):249-268

[14] Hastie T, Tibshirani R, Friedman J. The Elements of Statistical Learning - Data Mining, Inference, and Prediction. Springer Series in Statistics. Switzerland: Springer. 2009;**2**(1):93-85

[15] Ho TK. A data complexity analysis of comparative advantages of decision forest constructors. Pattern Analysis & Applications. 2002;**5**(2):105-112

[16] Friedman JH. Stochastic gradient boosting. Computational Statistics and Data Analysis. 2002;**38**(4):367-378

[17] Friedman JH. Greedy function approximation: A gradient boosting machine. The Annals of Statistics. 2001;**29**(5):1189-1232

[18] Mason L, Baxter J, Bartlett P, Frean M. Boosting algorithms as gradient descent in function space. NIPS Conference Proceedings. 1999:512-518

[19] Libbrecht MW, Noble WS. Machine learning applications in genetics and genomics. Nature Reviews. Genetics. 2015;**16**(6):321-332

[20] Längkvist M, Karlsson L, Loutfi A. A review of unsupervised feature learning and deep learning for time-series modeling. Pattern Recognition Letters. 2014;**42**(1):11-24

[21] Lloyd SP. Least squares quantization in PCM. IEEE Transactions on Information Theory. 1982;**28**(2):129-137

[22] Schmidhuber J. Deep learning in neural networks: An overview. Neural Networks. 2015;**61**(1):85-117

[23] Hazan E, Klivans A, Yuan Y. Hyperparameter optimization: A spectral approach. In: In: 6th International Conference on Learning Representations, ICLR 2018; Conference Track Proceedings. 2018

[24] Bardenet R, Brendel M, Kégl B, Sebag M. Collaborative hyperparameter tuning. In: 30th International Conference on Machine Learning; ICML. 2013. p. 2013

[25] Hutter F, Hoos H, Leyton-Brown K. An efficient approach for assessing hyperparameter importance. In: 31st International Conference on Machine Learning; ICML. 2014. p. 2014

[26] Hawkins DM. The problem of Overfitting. Journal of Chemical Information and Computer Sciences. 2004;**44**(1):1-12

[27] Ng A. Regularization—Solving the Problem of Overfitting. Coursera; 2011. Available from: https://www.coursera.org/learn/machine-learning/lecture/ACpTQ/the-problem-of-overfitting

[28] Srivastava N, Hinton G, Krizhevsky A, Sutskever I, Salakhutdinov R. Dropout: A simple way to prevent neural networks from overfitting. Journal of Machine Learning Research. 2014;**15**(56):1929-1958

[29] Picard RR, Cook RD. Cross-validation of regression models. Journal of the American Statistical Association. 1984;**79**(387):575-583

[30] Jothi N, Rashid NA, Husain W. Data mining in healthcare—A review. Procedia Computer Science. 2015;**72**(1):306-313

[31] Koh HC, Tan G. Data mining applications in healthcare. Journal of Healthcare Information Management. 2005;**19**(2):64-72

[32] Cottle M, Hoover W, Kanwal S, Kohn M, Strome T, Treister NW. Transforming Health Care through Big Data: Strategies for Leveraging Big Data in the Health Care Industry. Institute for Health Technology Transformation. New York, iHT2; 2013

[33] Wang Y, Kung LA, Byrd TA. Big data analytics: Understanding its capabilities and potential benefits for healthcare organizations. Technological Forecasting and Social Change. 2018;**126**(1):3-13

[34] Raghupathi W, Raghupathi V. Big data analytics in healthcare: Promise and potential. Health Information Science and Systems. 2014;**2**(1):3

[35] The Cancer Genome Atlas Program—National Cancer Institute [Internet]. 2019. Available from: https://www.cancer.gov/about-nci/organization/ccg/research/structural-genomics/tcga

[36] tcga-infographic-enlarge.__v100169753.png (1400×2580) [Internet]. 2019. Available from: https://www.cancer.gov/PublishedContent/Images/images/nci/organization/tcga/tcga-infographic-enlarge.__v100169753.png

[37] Peakman TC, Elliott P. The UK biobank sample handling and storage validation studies. International Journal of Epidemiology. 2008;**37**(2):234-244

[38] Sudlow C, Gallacher J, Allen N, Beral V, Burton P, Danesh J, et al. UK biobank: An open access resource for identifying the causes of a wide range of complex diseases of middle and old age. PLoS Medicine. 2015;**12**(3)

[39] Protocol for the UK Biobank—Wayback Machine [Internet]. 2019. Available from: https://web.archive.

org/web/20060214144838/http://www.ukbiobank.ac.uk/docs/draft_protocol.pdf

[40] Elliott P, Peakman TC. The UK biobank sample handling and storage protocol for the collection, processing and archiving of human blood and urine. International Journal of Epidemiology. 2008;**37**(2):234-244

[41] Gao GF, Parker JS, Reynolds SM, Silva TC, Wang LB, Zhou W, et al. Before and after: Comparison of legacy and harmonized TCGA genomic data commons data. Cell Systems. 2019;**9**(1):24-34.e10

[42] Bycroft C, Freeman C, Petkova D, Band G, Elliott LT, Sharp K, et al. The UK biobank resource with deep phenotyping and genomic data. Nature. 2018;**562**(7726):203-209

[43] Background—EPIC [Internet]. 2019. Available from: http://epic.iarc.fr/about/background.php

[44] Leitsalu L, Haller T, Esko T, Tammesoo ML, Alavere H, Snieder H, et al. Cohort profile: Estonian biobank of the Estonian genome center, university of Tartu. International Journal of Epidemiology. 2015;**44**(4):1137-1147

[45] Master Decoder: A Profile of Kári Stefánsson | The Scientist Magazine® [Internet]. 2019. Available from: https://www.the-scientist.com/profile/master-decoder--a-profile-of-kri-stefnsson-65517

[46] Gulcher J, Stefansson K. Population genomics: Laying the groundwork for genetic disease modeling and targeting. Clinical Chemistry and Laboratory Medicine. 1998;**36**(8):523-527

[47] China Kadoorie Biobank [Internet]. 2019. Available from: https://www.ckbiobank.org/site/

[48] Littlejohns TJ, Sudlow C, Allen NE, Collins R. UK biobank: Opportunities for cardiovascular research. European Heart Journal. 2017;**44**:1158-1166

[49] Scholtens S, Smidt N, Swertz MA, Bakker SJL, Dotinga A, Vonk JM, et al. Cohort profile: LifeLines, a three-generation cohort study and biobank. International Journal of Epidemiology. 2015;**44**(4):1172-1180

[50] The Health 202: NIH wants 1 million Americans to contribute to new pool of gene data. The Washington Post [Internet]. 2019. Available from: https://www.washingtonpost.com/news/powerpost/paloma/the-health-202/2018/01/16/the-health-202-nih-wants-1-million-americans-to-contribute-to-new-pool-of-gene-data/5a5ba45a30fb0469e8840135/

[51] FACT SHEET: President Obama's Precision Medicine Initiative. whitehouse.gov [Internet]. 2019. Available from: https://obamawhitehouse.archives.gov/the-press-office/2015/01/30/fact-sheet-president-obama-s-precision-medicine-initiative

[52] Precision Medicine Initiative (PMI) Working Group. The precision medicine initiative cohort program—Building a research foundation for 21st century medicine. Precision Medicine Initiative Work Group Report to Advisory Committee to Director NIH; 2015

[53] FinnGen, a global research project focusing on genome data of 500,000 Finns, launched. EurekAlert! Science News [Internet]. 2019. Available from: https://www.eurekalert.org/pub_releases/2017-12/uoh-fag121917.php

[54] Le QV. Building high-level features using large scale unsupervised learning. In: ICASSP, IEEE International Conference on Acoustics, Speech and Signal Processing—Proceedings. 2013

[55] Greenspan H, Van Ginneken B, Summers RM. Guest editorial deep

learning in medical imaging: Overview and future promise of an exciting new technique. IEEE Transactions on Medical Imaging. 2016;**35**(5):1153-1159

[56] Giger ML. Machine learning in medical imaging. Journal of the American College of Radiology. 2018;**15**(3):512-520

[57] Foster KR, Koprowski R, Skufca JD. Machine learning, medical diagnosis, and biomedical engineering research—Commentary. BioMedical Engineering Online. 2014. Online: Published 5 July 2014. Article number: 94

[58] Lundervold AS, Lundervold A. An overview of deep learning in medical imaging focusing on MRI. Zeitschrift fur Medizinische Physik. 2019;**29**(2):102-127

[59] Havaei M, Davy A, Warde-Farley D, Biard A, Courville A, Bengio Y, et al. Brain tumor segmentation with deep neural networks. Medical Image Analysis. 2017;**35**(1):18-31

[60] Islam J, Zhang Y. Brain MRI analysis for Alzheimer's disease diagnosis using an ensemble system of deep convolutional neural networks. Brain Informatics. 2018;**5**(2)

[61] Liu M, Zhang J, Adeli E, Shen D. Landmark-based deep multi-instance learning for brain disease diagnosis. Medical Image Analysis. 2018;**43**(1):157-168

[62] Demner-Fushman D, Chapman WW, McDonald CJ. What can natural language processing do for clinical decision support? Journal of Biomedical Informatics. 2009;**42**(5):760-772

[63] Kuleshov V, Ding J, Vo C, Hancock B, Ratner A, Li Y, et al. A machine-compiled database of genome-wide association studies. Nature Communications. 2019;**10**(1):3341

[64] Bedi G, Carrillo F, Cecchi GA, Slezak DF, Sigman M, Mota NB, et al. Automated analysis of free speech predicts psychosis onset in high-risk youths. NPJ Schizophrenia. 2015;**1**(1):15030

[65] Chang EK, Christine YY, Clarke R, Hackbarth A, Sanders T, Esrailian E, et al. Defining a patient population with cirrhosis: An automated algorithm with natural language processing. Journal of Clinical Gastroenterology. 2016;**50**(10):889-894

[66] Osborne JD, Wyatt M, Westfall AO, Willig J, Bethard S, Gordon G. Efficient identification of nationally mandated reportable cancer cases using natural language processing and machine learning. Journal of the American Medical Informatics Association. 2016;**23**(6):1077-1084

[67] Collins FS, Lander ES, Rogers J, Waterson RH. Finishing the euchromatic sequence of the human genome. Nature. 2004;**431**(7011):931-945

[68] Willyard C. Expanded human gene tally reignites debate. Nature. 2018;**558**. Online: Published 19 June 2018

[69] Ezkurdia I, Juan D, Rodriguez JM, Frankish A, Diekhans M, Harrow J, et al. Multiple evidence strands suggest that there may be as few as 19000 human protein-coding genes. Human Molecular Genetics. 2014;**23**(22):5866-5878

[70] Galvan A, Ioannidis JPA, Dragani TA. Beyond genome-wide association studies: Genetic heterogeneity and individual predisposition to cancer. Trends in Genetics. 2010;**26**(3):132-141

[71] Insel TR. Brain somatic mutations: The dark matter of psychiatric genetics. Molecular Psychiatry. 2014;**19**(2):156-158

[72] Diederichs S, Bartsch L, Berkmann JC, Fröse K, Heitmann J, Hoppe C, et al. The dark matter of the cancer genome: Aberrations in regulatory elements, untranslated regions, splice sites, non-coding RNA and synonymous mutations. EMBO Molecular Medicine. 2016;**8**(5):442-457

[73] Eichler EE, Flint J, Gibson G, Kong A, Leal SM, Moore JH, et al. Missing heritability and strategies for finding the underlying causes of complex disease. Nature Reviews. Genetics. 2010;**11**(6):446-450

[74] Zuk O, Hechter E, Sunyaev SR, Lander ES. The mystery of missing heritability: Genetic interactions create phantom heritability. Proceedings of the National Academy of Sciences of the United States of America. 2012;**109**(4):1193-1198. DOI: 10.1073/pnas.1119675109

[75] Gibson G, Dworkin I. Uncovering cryptic genetic variation. Nature Reviews. Genetics. 2004;**5**(9):681-690

[76] Kiser DP, Rivero O, Lesch KP. Annual research review: The (epi) genetics of neurodevelopmental disorders in the era of whole-genome sequencing—Unveiling the dark matter. The Journal of Child Psychology and Psychiatry and Allied Disciplines. 2015;**56**(3):278-295

[77] Fanciulli M, Petretto E, Aitman TJ. Gene copy number variation and common human disease. Clinical Genetics. 2010;**77**(3):201-213

[78] Park RW, Kim TM, Kasif S, Park PJ. Identification of rare germline copy number variations over-represented in five human cancer types. Molecular Cancer. 2015;**14**(25):1

[79] Zack TI, Schumacher SE, Carter SL, Cherniack AD, Saksena G, Tabak B, et al. Pan-cancer patterns of somatic copy number alteration. Nature Genetics. 2013;**45**(10):1134-1140

[80] Campbell CD, Eichler EE. Properties and rates of germline mutations in humans. Trends in Genetics. 2013;**29**(10):575-584

[81] Kuusisto KM, Akinrinade O, Vihinen M, Kankuri-Tammilehto M, Laasanen SL, Schleutker J. Copy number variation analysis in familial BRCA1/2-negative Finnish breast and ovarian cancer. PLoS One. 2013;**8**(8):e71802

[82] Zhang J, Walsh MF, Wu G, Edmonson MN, Gruber TA, Easton J, et al. Germline mutations in predisposition genes in pediatric cancer. The New England Journal of Medicine. 2015;**373**(24):2336-2346

[83] Redon R, Ishikawa S, Fitch KR, Feuk L, Perry GH, Andrews TD, et al. Global variation in copy number in the human genome. Nature. 2006;**444**(7118):444-454

[84] Auton A, Abecasis GR, Altshuler DM, Durbin RM, Bentley DR, Chakravarti A, et al. A global reference for human genetic variation. Nature. 2015;**526**(7571):68-74

[85] Bredfeldt TG, Walker CL. Epigenetics. In: Comprehensive Toxicology. 2nd ed. 2010

[86] Heard E, Martienssen RA. Transgenerational epigenetic inheritance: Myths and mechanisms. Cell. 2014;**157**(1):95-109

[87] Toh C, Brody JP. Chromosomal scale length variation of germline DNA can predict individual cancer risk. bioRxiv. 2018;**10**(1101):303339. DOI: 10.1101/303339

[88] Lello L, Raben T, Yong SY, Tellier LC, Hsu SDH. Genomic prediction of complex disease risk. bioRxiv. 2018;**10**(1101):506600. DOI: 10.1101/506600

Chapter 5

Radio Systems and Computing at the Edge for IoT Sensor Nodes

Malcolm H. Smith

Abstract

Many Internet of Things (IoT) applications use wireless links to communicate data back. Wireless system performance limits data rates. This data rate limit is what ultimately drives the location of computing resources—on the edge or in the cloud. To understand the limits of performance, it is instructive to look at the evolution of cellular and other radio systems. The emphasis will be on the RF front-end architectures and requirements as well as the modulation schemes used. Wireless sensor nodes will often need to run off batteries and be low-cost, and this will constrain the choice of wireless communications system. Generally cheap and power efficient radio front ends will not support high data rates which will mean that more computing will need to move to the edge. We will look at some examples to understand the choice of radio system for communication. We will also consider the use of radio in the sensor itself with a radar sensor system.

Keywords: IoT sensor nodes, wireless systems, 5G, cellular, WLAN, Bluetooth, radar, RF front end, radio circuitry, bandwidth, power requirements

1. Introduction

Many IoT systems consist of networks of sensors, the data from which is brought together and processed to give some desired results. The method used to connect the sensor to the Internet to enable the transfer of data is a design factor that needs to be considered early on. It is not the case that throwing a lot of bandwidth at the problem and processing the data together in a central location are the solution in most cases. The answer is do not "use 5G" for all systems!

There are many trade-offs involved in the specification of a radio system to be used for Internet of Things (IoT). The main trade-off is the bitrate and power trade-off. Sending messages at a high bitrate requires more power than that at a low bitrate. There is also a cost and bandwidth trade-off: high-bitrate solutions require a wide bandwidth which, generally, costs more than the low bandwidth. Finally, there is the often-overlooked fact that high bitrate solutions require high fidelity of the radio which means more receive power. This limits their use to applications where the separation of transmitters and receivers is small.

If cheap high-bitrate wireless communications are not available, what are the alternatives if the sensor needs high bandwidth (video cameras being the most obvious example)? The most obvious solution is to use a wired communication link. This may not be a viable option for many IoT solutions where the access to wired links is not available. The alternative is to use computing at the edge.

By processing the signal at source—computing at the edge—we can decrease the bandwidth requirement, and, hence, we can use an alternative wireless communication solution.

First, we need to review digital communications as they pertain to wireless systems [1].

2. Review of digital communications systems

The wireless systems used in IoT applications will be digital systems. Although the world they are interfacing to is analogue, IoT systems will have a means of converting those analogue signals that come from the analogue world through sensors to digital signals. These digital signals will be processed ultimately in digital computers whether it be on the edge or in the cloud. It is, therefore, instructive to review digital communications systems.

2.1 Introduction

Modern digital communications systems are complex with many layers that interact with each other. In this chapter we are only going to consider the lowest layer of the Open Systems Interconnection (OSI) model, the physical layer. Furthermore we are going to consider this layer from the perspective of the analogue and RF components of the system as they are the parts that drive a lot of the trade-offs for power consumption that will be a major driving factor in the selection of a radio system for IoT.

2.2 Complex signals

In order to understand the concepts of digital communications systems and even the hardware implementation of them, it is necessary to understand complex signaling. We will therefore review the concept of complex signaling and in-phase and quadrature (IQ) modulation.

A complex signal can be represented as a function of time using a complex exponential:

$$s(t) = Ae^{j2\pi ft} \tag{1}$$

Expanding this exponential using Euler's formula gives us this complex exponential function represented using trigonometric functions:

$$s(t) = A\cos(2\pi ft) + j \times A\sin(2\pi ft) \tag{2}$$

where j is $\sqrt{(-1)}$, A is a scaler quantity representing the magnitude of the signal, and f is a scalar representing the frequency in Hz (or cycles per second).

Graphically this can be represented by a vector from the origin with a magnitude of A and an angle relative to the positive real axis of $2\pi ft$. This vector will spin around the origin in an anti-clockwise direction tracing out a circle. The number of times it spins in 1 second is given by the value of f, so for a value of 1 Hz (1 cycle per second), the vector will complete one rotation.

If we introduce another vector with a negative value for the frequency:

$$s(t) = Ae^{-j2\pi ft} = A\cos(-2\pi ft) + j \times A\sin(-2\pi ft) \tag{3}$$

we get a second vector that rotates in a clockwise direction at the same rate and tracing out the same circle. From the trigonometric identities:

$$\cos -\theta = \cos\theta \tag{4}$$

$$\sin -\theta = -\sin\theta \tag{5}$$

we can see that if we sum the two vectors, the imaginary components are canceled and we are left with a real cosine with twice the amplitude of the vector components. This is illustrated in **Figure 1**.

By shifting the vectors by 90°, we get a sine with twice the amplitude of the vector components. This leads us to the basic formulae relating real sinusoidal functions to complex exponential functions:

$$A\ \cos(2\pi ft) = A \times \frac{e^{j2\pi ft} + e^{-j2\pi ft}}{2} \tag{6}$$

$$A\ \sin(2\pi ft) = A \times \frac{e^{j2\pi ft} - e^{-j2\pi ft}}{2 \times j} \tag{7}$$

From these equations real signals separated by 90° can be used to generate complex signals. Signals at 90° such as this are referred to as quadrature signals and the individual component signals as in-phase (I) and quadrature (Q). These signals will be at low frequency and sit either side of 0 Hz (viz. DC)—they are what we call baseband signals. We need a method of transferring these baseband signals to radio frequency (RF) or microwave frequencies.

Using trigonometric identities, we can prove that:

$$A(t)\ \sin(2\pi ft + \theta(t)) = A(t) \times (\sin 2\pi ft \times \cos\theta(t) + \cos 2\pi ft \times \sin\theta(t)) \tag{8}$$

What this tells us is that we can upconvert a baseband signal represented by:

$$I_{BB} = A(t) \times \cos\theta(t) \tag{9}$$

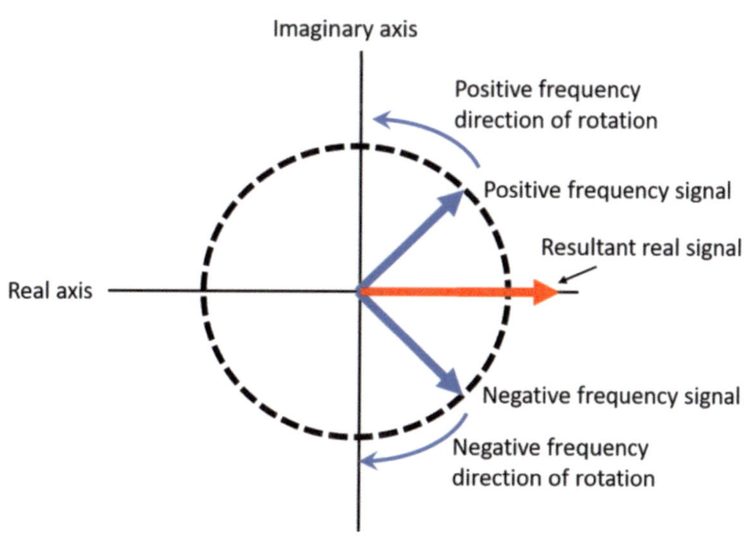

Figure 1.
Real cosine signal as a vector sum of two complex signals: A positive frequency signal rotating anti-clockwise and a negative frequency signal rotating clockwise.

$$Q_{BB} = A(t) \times \sin\theta(t) \tag{10}$$

with another signal, known as a local oscillator (LO) signal, represented by:

$$I_{\text{LO}} = \sin 2\pi ft \tag{11}$$

$$Q_{\text{LO}} = \sin\left(2\pi ft + \frac{\pi}{2}\right) = \cos\left(2\pi ft\right) \tag{12}$$

We can implement Eqs. (8)–(12) in circuitry using circuits that can multiply two real signals (circuits of this sort are known as mixers) by signals generated from an oscillator with phase shifters (LO generator) and some sort of circuit to sum the resultant outputs (usually done by summing currents). Such a circuit is called a quadrature modulator or quadrature mixer, and the figurative block diagram is shown in **Figure 2** below with the local oscillator generation.

The receive operation is just the reverse of the above—the RF signal is split into *I* and *Q* components using a quadrature mixer driven from a quadrature LO signal. This is illustrated in **Figure 3**.

2.3 Encryption

On top of all the other parts of the data transmission system is the need to encrypt the data. This needs to happen for the data itself and for any control

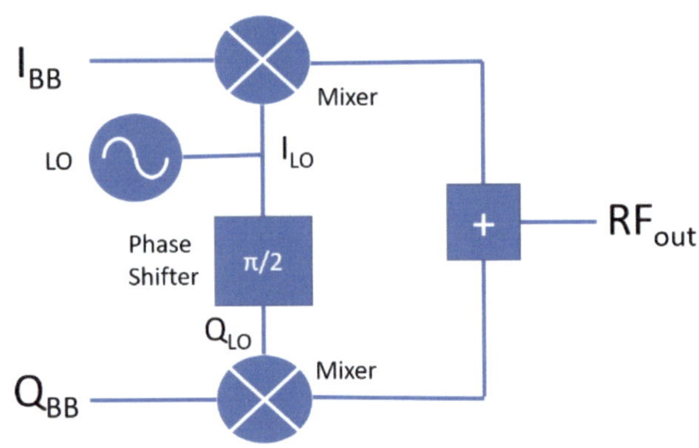

Figure 2.
Quadrature upconverter with local oscillator (LO) and quadrature local oscillator generation.

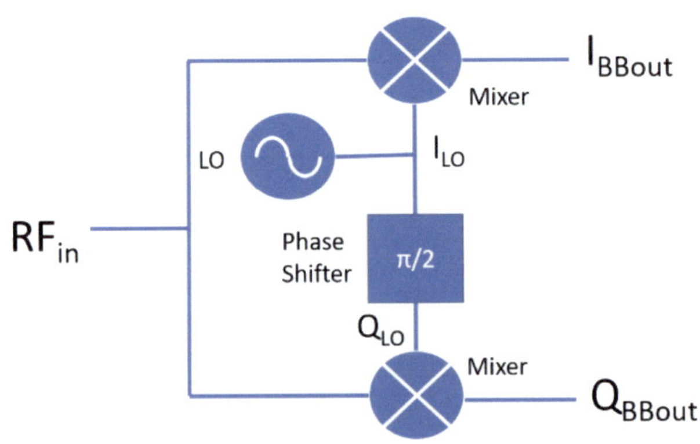

Figure 3.
Quadrature downconverter with local oscillator (LO) and quadrature local oscillator generation.

channels. It is obviously important to encrypt the data itself as this is what has value for the user. However, if control channels aren't encrypted, then an outside agent can read the control information and even take over the channel. This means it is necessary to encrypt control channels as well. Encryption will generally not add much to the overall size of data being sent and so does not affect the efficiency by sending more data. However, the computations involved in encryption do require extra power and so do lower efficiency.

2.4 Interleaving

When a radio receiver is moving, it goes through fades in signal strength due to two or more reflections of the signal interfering. This potentially leaves holes in the data stream, and these holes can lead to loss of information if related bits are sent together. In order to avoid this, the data bits can be separated so that consecutive bits are not sent together. An analogy would be a deck of cards arranged by suit. If four cards are taken together from a single place at random, it could mean, for instance, that the jack, queen, king, and ace of spades are all removed. If the pack is shuffled first, then the four cards taken from adjacent positions will not be related, and arranging the cards in suits, we should be able to spot the pattern even if individual cards are missing. Interleaving is like the shuffling of the pack in the example; however it is done in a controlled manner so that the data can be reconstructed.

2.5 Error coding and correction

Transmission of data will result in errors due to various random processes in the transmission. It is important to be able to a first level detect that an error has occurred and that the data is corrupted. It is much better to be able to correct errors that occur. The simplest method for correcting errors is to have the transmitter retransmit the data if an error is detected or no message acknowledging receipt of the data is received by the transmitter. This method is known as automatic repeat request (ARQ). An improved form of error correction, known as forward error correction, is achieved by adding extra bits to the data that is transmitted to enable both detection and correction of this data. Examples of error correction codes include convolutional codes that are added and processed on a bit-by-bit basis and Reed-Solomon codes or turbo codes that are added and processed on a block-by-block basis. If an error is detected in forward error correction that cannot be corrected, then ARQ can still be used and the data retransmitted. Adding error correction codes adds extra data to the transmission and so decreases the overall power efficiency of the system.

2.6 Modulation

Transmitting data as a series of "0's and 1's" over the transmission medium would require a very wide bandwidth. The required bandwidth is a function of the rise and fall time of the data rather than the clock rate and is significantly higher than the clock rate of the data. To transmit over air for a wireless system, the data needs to be bandlimited first. A filter must be used to bandlimit the data, but the performance of the filter is important. The filter needs to be sharp enough to filter the data to fit in an assigned bandwidth (radio channel) but not destroy the higher frequencies of the data, and so it must have a sharp cut-off. The time domain performance (impulse response) of the filter is also important. If a signal is bandlimited in the frequency domain, then it tends to spread out in the time domain. This is obviously undesirable with a digital signal as one data symbol would interfere with subsequent symbols. There are two approaches to the impulse response requirement:

1. Allow inter-symbol interference to happen, and correct or allow for it in the receiver. In this case the receiver will contain an equalizer to provide an inverse filter to the original filter (and any filtering from the channel as well). Alternatively, the receiver may implement the Viterbi algorithm to decode the data signal.

2. Use a filtering scheme; a root-raised cosine (RRC) filter is common, that has nulls at the subsequent symbols. In this case the filter on the transmit is matched with an identical filter in the receiver. In the case of an RRC filter, the impulse is a sinc function which naturally has nulls at integer time intervals. A data signal passed through such a filter will look like a sequence of overlapping sinc functions with the nulls of the previous sinc functions occurring at the peaks of the subsequent ones. This is illustrated in **Figure 4**.

To transfer a digital bit stream over an analogue channel, it is necessary to use some form of modulation. In modulation, a periodic signal, carrier signal, has its frequency, its amplitude, or phase or both changed—modulated—by a second signal. For digital modulation, this second signal is a filtered version of the digital bit stream. Modulation schemes can be divided into many different groupings—phase modulation, frequency modulation, and amplitude modulation—but for the front end, the grouping that matters is non-linear also known as constant amplitude modulation (AM) and linear modulation. The bit stream is broken up into symbols. Each symbol can be one, two, three, or more bits long. These symbols are used to modulate the carrier. To send a data stream over a link with a given bandwidth, in general a symbol with more bits is needed. This will mean that both amplitude and phase will need to be modulated and the system will be more susceptible to errors.

The simplest modulation to visualize is amplitude modulation (AM). The amplitude of the carrier wave is modulated by the modulating signal. We are familiar with AM radio where the signal doing the modulation is an analogue signal. This modulation can be demodulated by using a simple diode, and in the simplest receivers, this diode was made from a crystal (lead sulphide) with a "cat's whisker' touching it leading to the term "cat's whisker radio" or "crystal radio" [2]. For digital bit-streams, "On–off keying" is the simplest form of amplitude modulation

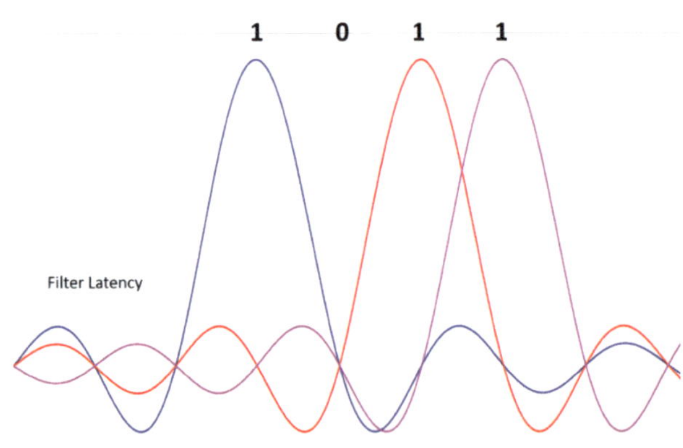

Figure 4.
A filter with a sinc impulse response being fed the bit sequence (1011). The filter latency means that the signals will be delayed with respect to the input.

and is known as amplitude shift keying (ASK). In practice the carrier is usually not switched on and off as this causes spurious signals around the transition but is switched between two amplitudes. A signal that is switched sharply between two different amplitudes is wideband, so a filtered signal is used leading to a smooth transition between the two signal levels (**Figure 5**).

In frequency modulation (FM), the frequency of the carrier is modulated. Frequency modulation is the FM in FM radio, and in this case the modulating signal is analogue [3]. The simplest example of FM use as a digital modulation would be using on–off keying where the carrier is switched between two frequencies. This is known as frequency shift keying (FSK). If more than two frequencies are used, it is possible to send more than one bit at a time—for instance, three bits would require eight different frequencies (**Figure 6**).

Again, to limit the bandwidth, the input signal is filtered, and the transition passes through the intermediate frequencies. A typical version of filtered FSK is GFSK which uses a Gaussian filter before to limit the bandwidth—GFSK is used in many radio systems, Bluetooth being a good example. An efficient form of

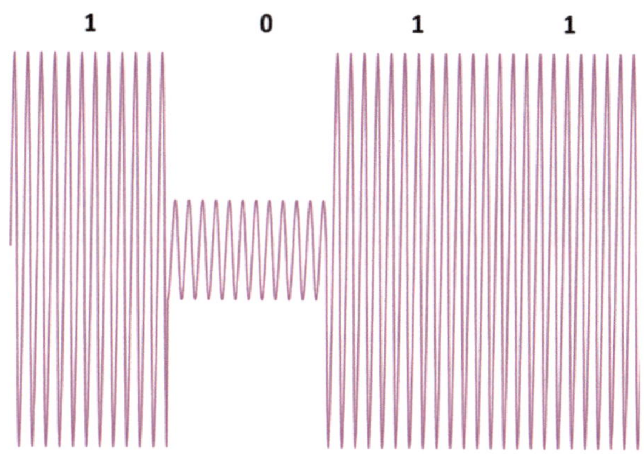

Figure 5.
Amplitude shift keying (ASK) with two levels (no filtering) modulated by the sequence [1011].

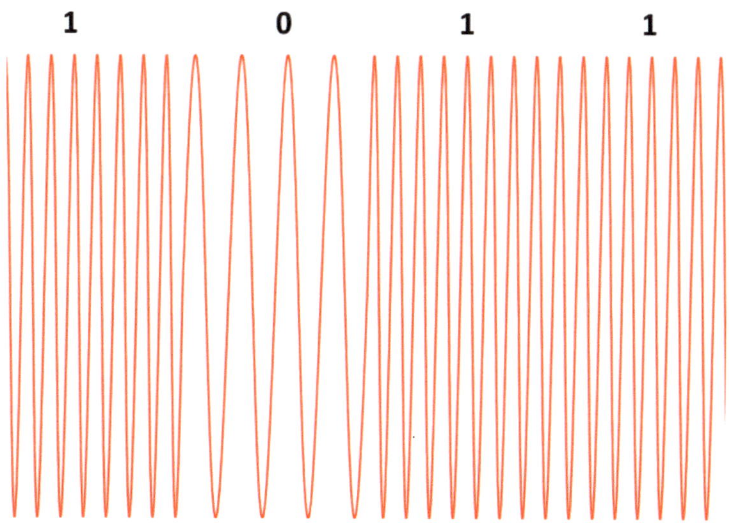

Figure 6.
Frequency shift keying (FSK) with two levels (no filtering) modulated by the sequence (1011).

frequency modulation is minimum shift keying (MSK). In MSK, the two frequencies used differ by half the bitrate, and this gives a very efficient modulation with a modulation index of 0.5. If Gaussian filtering is used to limit the bandwidth with MSK, the resulting modulation is known as GMSK, and this is the modulation scheme used in GSM. In FM radios (digital and analogue), the amplitude of the signal is not important, and so the radio is much more robust and tolerant of signal fading. Therefore, FM radio was the preferred medium for broadcast radio before the arrival of digital radio. However, FM requires a more complicated receiver than AM. As the amplitude is not important—it is a non-linear modulation—the transmitter can be simplified vs. a transmitter used for modulations where the amplitude varies (linear modulations).

In phase modulation (PM) schemes, the phase of the carrier is varied. Again, on–off keying can be used to vary the phase, such a modulation being known as phase shift keying (PSK). The simplest form is binary phase shift keying (BPSK) where only one bit is used, and the phase is varied by 180° to represent a binary "0" or "1". In theory, BPSK, like all pure phase modulations, is a constant envelope modulation (there is no change in the amplitude), but that would require an infinite bandwidth to accommodate an instantaneous 180° phase shift. In practice a BPSK where the bandwidth is limited will have amplitude modulation associated with it. Both unfiltered and filtered BPSK are shown in **Figure 7**. BPSK is the modulation used in Zigbee for the low band.

If a two-bit symbol is to be transmitted, then four phases must be used each 90° apart; such a modulation is known as quadrature phase shift keying (QPSK), and variants are used in many commercial radio systems. For a three-bit symbol, eight phases are needed each 45° apart, and this system is known as 8PSK. The constellation points for QPSK and 8PSK are shown in **Figure 8**. More bits can be added to the symbol, and the number of phases increased for each extra bit by a factor of two, but this rapidly becomes impractical as the tolerance of error decreases as each extra bit decreases the distance between phases by a factor of two. QPSK, 8PSK, and higher orders of PSK have an amplitude component and require linear transmitters. This also means that the power amplifier needs to be linear and so is less efficient than the switching power amplifiers used in non-linear modulations like GMSK or GFSK.

For amplitude modulation and frequency modulation systems, it is possible to consider only one scalar quantity (amplitude or frequency). For phase-modulated systems, we must look at amplitude and phase, and that requires us to look at a complex representation of the signal at baseband (real, imaginary). We can map the signal out on the complex plane and see what it looks like. For a QPSK signal, the four decision points are 90° apart. We can easily visualize the points (1,0), (0,1), (−1,0), and (0, −1). However, as long as the four points are 90° apart, the location

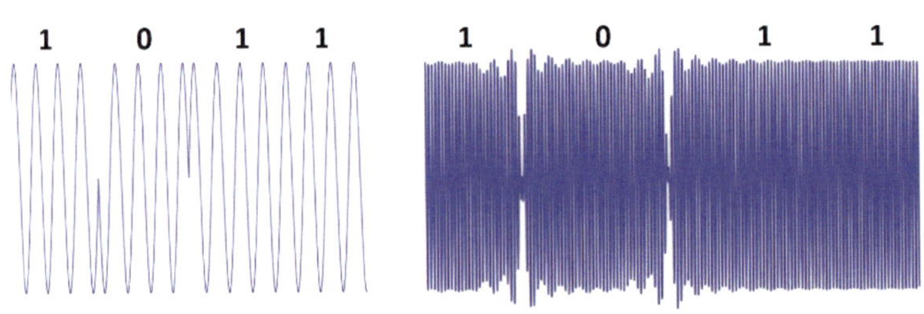

Figure 7.
Binary phase shift keying (BPSK). The left-hand side is without filtering on the bit stream which has shown modulating a low-frequency signal to make the transitions clearer; the right-hand side is with filtering of the data stream but with a higher frequency to make the resulting amplitude modulation clearer.

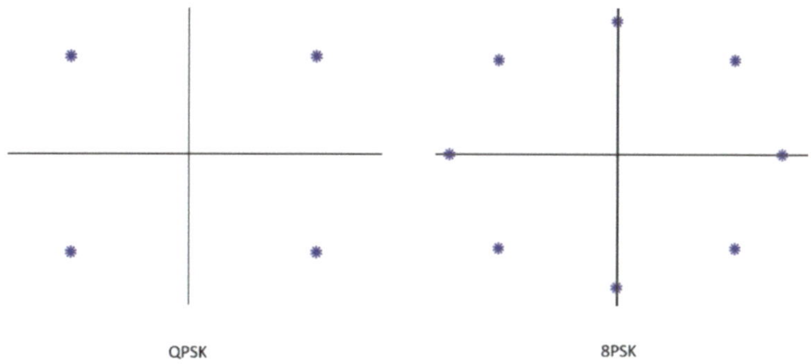

Figure 8.
The QPSK and 8PSK constellation points plotted in the complex domain.

of the initial point can be at any random phase around the circle, and so a 45° phase shift also gives valid points at ($\sqrt{2}$, $\sqrt{2}$), ($-\sqrt{2}$, $\sqrt{2}$), ($-\sqrt{2}$, $-\sqrt{2}$), and ($\sqrt{2}$, $-\sqrt{2}$). This arrangement always includes a path that passes through the origin—in other words the amplitude goes to zero—and this is bad for the power amplifier as it must be sufficiently linear to be able to handle this change through zero amplitude.

One solution to the need to pass through zero is to offset one of the in-phase and quadrature bit streams by half a clock cycle ensuring that the code points never change by more than 90°, so the modulation will never pass through zero. This scheme is known as offset QPSK (OQPSK). Another solution is to rotate the constellation of points by 45° at each symbol so that there is no path through the origin and the amplitude never goes to zero. This is known as $\pi/4$ QPSK. Both these schemes are used in radio standards to simplify the requirements for the power amplifier (PA). The constellation diagrams of QPSK and $\pi/4$ QPSK are shown in **Figure 9**.

The final sort of modulation we will consider is quadrature amplitude modulation (QAM). QAM modulates both the phase and the amplitude. Generally, it uses points that are in a square centred on the origin, so QPSK is actually a form of QAM. This means that powers of two that are also square numbers are preferred numbers of constellation points: 4 QAM (QPSK), 16 QAM, 64 QAM, 256 QAM, 1024 QAM, etc. A constellation diagram for a 16 QAM implementation is shown in **Figure 10**.

QAM can support high data rates with a relatively small bandwidth, and this makes it desirable in applications where a high bitrate is desired. However, because the spacing between constellation points is less than in other modulations and

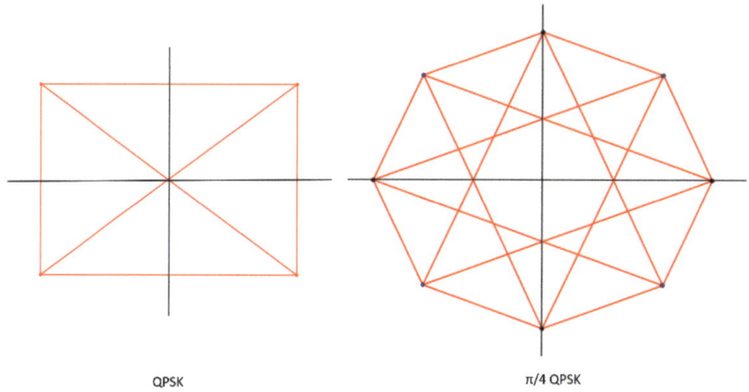

Figure 9.
QPSK and $\pi/4$ QPSK constellation diagrams. Note that the $\pi/4$ QPSK constellation does not pass through the origin simplifying the design of the PA.

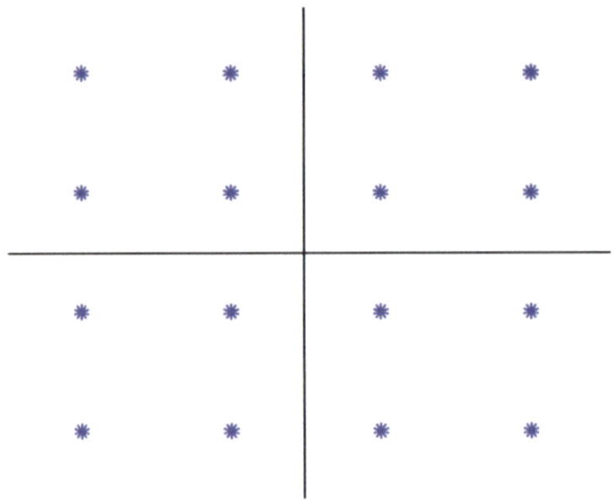

Figure 10.
Constellation points for 16 QAM modulation.

decreases as more points are added, QAM requires more precision in the transmit architecture and is less robust on the receive side to noise and fading. On the transmit side, the transmitter and the power amplifier need to be very linear. On the receive side, the whole receiver needs a much wider dynamic range than other modulations.

2.7 Channel access method

When sharing a communications medium, the users need to be allowed access to the medium in a controlled manner. There are several techniques that can be used to control access to the medium. Usually a combination of several of these methods is needed to get a system to work well.

The first, and perhaps most obvious, is spatial separation known as spatial division multiple access (SDMA). This is the principle used in the cellular network where an area is separated into cells with channel allocation such that neighboring cells do not share the same channels but channels are reused in cells that are sufficient far apart that they will not interfere with another.

The next technique is to separate the channels by frequency using a technique called frequency division multiple access (FDMA). We are all familiar with this from AM and FM radio stations as well as terrestrial broadcast television. In fact, all radios use this technique in some form or another; as radio signals of a certain type are restricted to certain bands, the access to most is controlled by a government agency. Within a radio band, users can be assigned channels that are also of different frequencies and are a sub-band of the overall radio band assigned to that system.

The next technique is time division multiple access (TDMA). In this case the resource is split into units of time generally called slots. A user is allocated a slot to transmit on and is quiet in other slots. There will generally be a matching slot (or slots) on the receive side to receive transmissions. TDMA is good for transmissions that are bursty in nature which many IoT applications are. TDMA is used in the GSM system.

The final technique is code division multiple access (CDMA). Each user is allocated a code that is unique to them and mathematically orthogonal to other user codes. The data stream is multiplied with a much faster version of the code and transmitted. On the receive side, the receiver uses the same code and again multiplies the incoming data by that code to decode the message. The principle is simple, but the actual implementation is complex because of the need to time align the code with the received signal and the presence of reflected copies of the signal.

2.8 Orthogonal frequency division multiplexing

Orthogonal frequency division multiplexing (OFDM) is a technique mainly used to transmit wideband data. OFDM has a number of advantages for the transmission of wideband data which is why it is used in all new wideband systems:

1. It is robust against narrow band interference from other radios.

2. It is robust against fading.

3. It deals with multi-path easily.

However, is also has disadvantages:

1. It is sensitive to Doppler shift.

2. It has a high peak-to-average ratio (PAR) which requires a linear power amplifier and a lot of current.

For IoT systems that aren't designed to be particularly mobile, the sensitivity to Doppler shift is probably not an issue. However the inefficiency of a good linear power amplifier will be a significant issue in systems where battery current is premium.

In an OFDM system, the radio band is split into a number of sub-bands with sub-carriers which are modulated separately. Each sub-carrier needs to be orthogonal to the others which gives the relationship:

$$\Delta f = \frac{1}{\mathrm{T_s}} \tag{13}$$

where Δf is the difference in the sub-carrier frequencies and T_s is the receive symbol duration. This gives an overall bandwidth of:

$$B \cong \Delta f \times N \tag{14}$$

where B is the bandwidth and N is the number of sub-carriers. Bandwidth efficiency is high as the orthogonal nature of the sub-carriers ensures that adjacent sub-bands do not interfere with the demodulation of a sub-band, so no guard bands are needed, and spectral efficiency is high (**Figure 11**).

Within each sub-band, the sub-carrier is modulated just as it would be in a single carrier system. QAM-based modulations are common starting with QPSK, although 802.11a allowed BPSK, and increasing the symbol size and modulation as the communications channel quality improves.

Usually the centre sub-carrier of the overall bandwidth is not modulated. This assists on in the receiver where it converts down to DC. Modern receivers are generally what is known as direct downconversion receivers, and the control of DC offsets is an issue. With no useful signal around DC, the DC offset can be eliminated using simple filtering schemes.

A variant of OFDM, orthogonal frequency division multiple access (OFDMA), shares the band between users rather than dedicating the whole band to a single user.

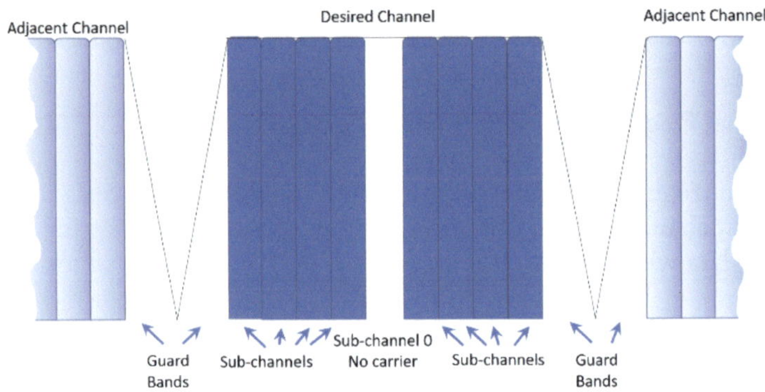

Figure 11.
OFDM channel structure.

This is done by designating each user a group of sub-carriers. It is used on the downlink of the cellular standards of the Long Term Evolution (LTE) family and 5G.

A variant of OFDM, single carrier frequency division multiple access (SC-FDMA), is used in the cellular standards on the uplink. In SC-FDMA, the bit stream is processed in a manner similar to OFDMA, but the parallel streams are then serialized to give a single carrier modulation. This gives a signal with much less PAR which requires a less linear power amplifier and transmit section and so saves power. As user equipment is battery driven and therefore sensitive to power consumption, this is necessary for the quality of service it can deliver customers.

2.9 Half- and full-duplex

How a radio system handles the relationship between transmit and receive has a large impact on the design of the radio front end and the power consumption of the overall radio. It is possible to transmit while receiving, and a lot of modern radio systems have this capability. These systems are known as full-duplex systems. Obviously, if they are to transmit while receiving, they cannot use the same frequency; otherwise they block themselves. (The exception here is radar which we will cover later.) The systems that do not permit transmission at the same time as reception are called half-duplex systems and are significantly simpler and more power efficient.

There are two approaches to duplexing: frequency division duplex (FDD) and time division duplex (TDD). FDD is the only approach that allows for full-duplex operation as the transmitter and receiver are operating at different frequencies. Usually this requires a front-end filter called a duplexing filter or duplexer that filters the transmit out on the receive path and filters transmit noise in the receive band out on the transmit path. Cellular radios usually separate the transmit and receive bands and so are capable of full-duplex operation even though some of the earlier standards did not call for it.

TDD is a half-duplex technique that is used a lot in connectivity radios (e.g. WiFi, Bluetooth) and is specified for some cellular radio standards but is not as common there. A TDD radio uses the same frequency for transmit and receive and will transmit and then switch to receive to listen for any reply.

3. Overview of radio transmit and receive architectures

Figure 12 is a block diagram showing the structure of a constant envelope transmitter. As the data is only coded in the phase, the amplitude contains no

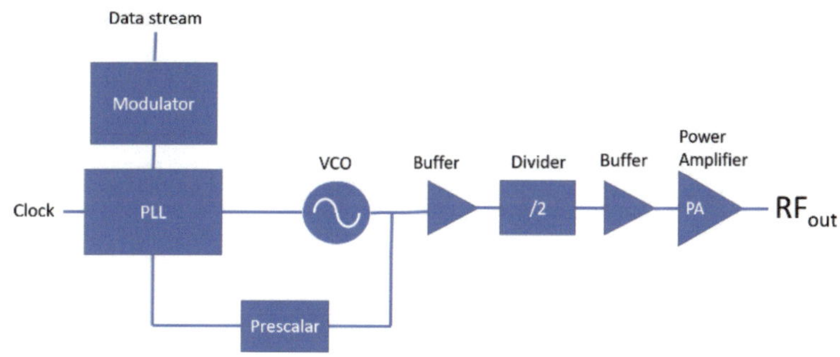

Figure 12.
Constant envelope transmit employing a PLL.

information. It is still necessary to control the amplitude in many radio systems to ensure that the signal received by the base station is not too large or too small. However, this amplitude control only needs to be accurate enough to ensure the signal is within a few dB of target and can be relaxed compared to a system that uses linear modulation.

With no requirement to provide modulation on the amplitude, only a circuit that modulates the phase is required, and a PLL is the best circuit for the job. The first transmitters of this type were introduced for GSM and used offset-loop PLLs to be compatible with the Cartesian baseband circuit outputs available at the time. These were quickly replaced by what are known as fractional N (frac-N) frequency synthesizers. In a frac-N frequency synthesizer, the divider is controlled by a sigma-delta modulator, and the divide ratio is constantly updated between integer values allowing the effective average ratio to be a fraction of the nominal integer divide ratio. This operation is carried out in the digital domain making circuit design easier.

The noise from the PLL and associated circuits is filtered by the PLL filter. This means that the noise far away from the carrier is dominated by the voltage-controlled oscillator (VCO), divider, and power amplifier (PA) noise. As noise away for the carrier is important in radio systems, this is a big advantage for these constant envelope transmitters. It generally means that the filtering at the output of the PA can be a simple harmonic trap filtering—filters out harmonics of the transmit signal—and does not need to filter close to the transmit signal. Harmonic trap filters can be implemented using standard passive components (inductors and capacitors) not specialized components (SAW, BAW, or FBAR filters), and these filters will generally have lower loss (meaning the overall transmitter is more efficient).

The design of the VCO is critical. It needs to have a very low noise profile as this noise will dominate the noise of the transmitter, and it needs to be able to drive the output well. Following the VCO is a divider. This is usually a divide by two dividers, so the VCO frequency is set at twice the desired output frequency. The use of a frequency of twice the transmit frequency is to stop feedback from the PA causing the VCO to shift off frequency. This effect is known as pulling and happens if the output signal is at the same frequency as the VCO frequency as the VCO will injection lock to the PA output signal. In some cases, even with a VCO being run at twice the transmit frequency, the VCO can still be pulled if the PA produces lots of second harmonic distortion. In this case a VCO frequency of four times the transmit frequency can be used with cascaded dividers that divide by two each. Running the VCO at twice or four times the transmit frequency obviously has power implications because the transistors will have less gain at higher frequencies. This is an unavoidable trade-off in these designs.

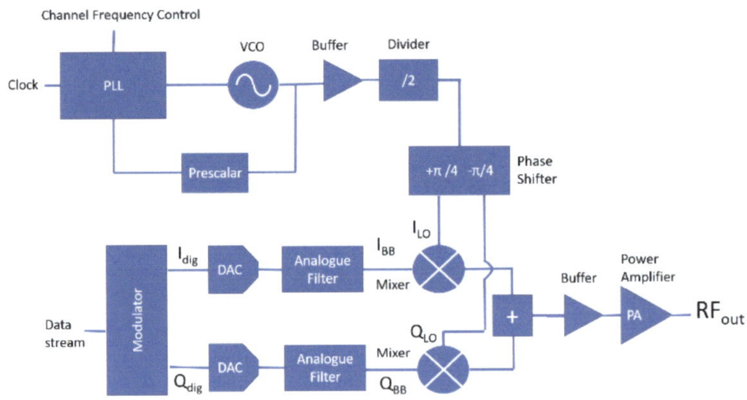

Figure 13.
Cartesian upconverter-based transmit for linear modulations with a PLL to control the local oscillator. Modulation is done on the digital domain giving digital I *and* Q *signals which are converted to analogue for upconversion.*

Figure 13 is a simplified block diagram of a typical transmitter used for cellular and connectivity solutions in smartphones. This sort of transmitter is known by many names, but *I* will refer to it as a Cartesian transmitter here. The baseband signal is coded as a complex signal with an in-phase (*I*) and quadrature (*Q*) component. This allows us to control both the amplitude and the phase and so is used for linear modulations where information is coded in both the amplitude and the phase.

It is obvious that this transmitter is a lot more complex than the constant envelope transmitter. In fact, the Cartesian transmitter uses most of the components of the constant envelope transmitter just to generate a constant local oscillator (LO) signal. The complexity not only means an increase in cost but also an increase in power consumption.

The linear modulation signal has both amplitude and phase components, and so the signal chain needs to be linear. This means it needs to have a constant DC bias current flowing and so is less efficient than a non-linear equivalent that can just operate as a digital logic gate switching between states. This inefficiency shows up particularly in the PA which can have half the efficiency of its non-linear counterpart.

One extra effect to consider for a transmitter suitable for linear modulations is the noise. It is a general rule that any component added to a system will add noise. Whether this added noise is large enough to affect the overall signal to noise of the system is somewhat a matter of design. In the case of the Cartesian transmitter, the signal path adds a substantial amount of noise to the signal being transmitted. In past the signal was usually cleaned up by adding a filter between the transmitter and the PA. This was easily achieved as the transmitter and the PA are usually on separate integrated circuits even now, and so the signal had to pass through the main circuit board to which a filter could be easily added. However, filters are expensive, and as the number of bands increased, the transmitter had to serve more bands which meant more filters, and some form of isolation was needed for unused bands. The filter was eliminated by cutting the noise of the transmit chain itself.

The VCO for the LO needs to be at a different frequencies from the transmit for the same reason as the constant envelope transmitter, namely, VCO pulling. If the VCO is not being operated at some integers multiple of the desired frequency, then extra circuits are needed to generate the desired frequency. The VCO can operate at twice the desired frequency, but if possible, four times makes more sense as it is easier to generate LO signals that are exactly 90° apart. Having LO in-phase and quadrature signals that are exactly 90° apart helps keep the number of errors down in the overall transmitter.

Figure 14 shows a simplified receiver. It looks somewhat like a Cartesian transmitter in reverse. The PA is replaced by a low noise amplifier (LNA), and again two LO signals that are 90° apart are used to downconvert to baseband. This scheme is called direct downconversion. In the past, heterodyne receivers were used, and the signal was converted down through one or more intermediate frequencies (IFs) before being brought down to the baseband. Consumer radio receivers, including those for the cellphone network where more performance is needed, are all direct downconversion receivers.

There are no pulling issues as such in the receiver, so the VCO for the LO can be at the same frequency as the receive signal, but running at twice or four times the frequency makes the generation of the quadrature components easier using a divided down VCO. For a full-duplex radio, it is possible to share the VCO between the receive and the transmit, but this requires a method to derive the transmit frequency and receive frequency from the same source. Two synthesizers can be used, but care must be taken to ensure they do not talk to each other.

On the receive side, as on the transmit side, most of the power consumption occurs in the RF blocks. These are the LNA, mixer, and LO buffer. The baseband filtering will not consume as much power, but the power consumption in the baseband filters is a function of the bandwidth of the signal: wider bandwidth leads to more power consumption. Having a narrower bandwidth not only lowers the power requirements for the filtering but also allows for more resolution in the ADC as it can run at a lower frequency than a wider bandwidth radio, and resolution and bandwidth tend to have a reciprocal relationship. With higher resolution on the ADC, it is possible to move more of the filtering and gain functions into the digital which makes the design easier, smaller, and often lower power.

Figure 15 shows the radio front end for a half-duplex radio (the transmit and receive never operate at the same time) and a full-duplex radio. Both use frequency division duplexing (as would be seen in a cellular radio system). Many elements overlap with the previous figures showing transmit and receive structures. This is so that you can see how everything fits together. In the cellular world, the front end started out as discrete components. Later, the LNAs were integrated into the transceiver chips, but the other RF components were kept separate and integrated into

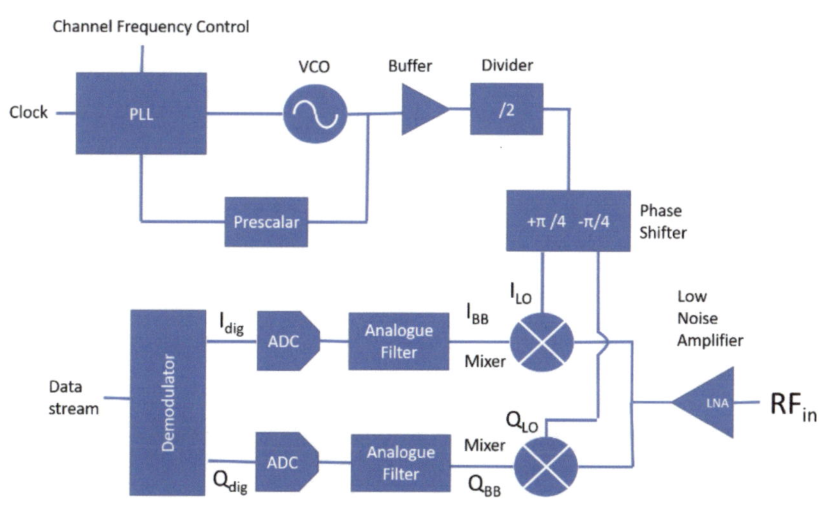

Figure 14.
Typical architecture for a direct downconversion receiver.

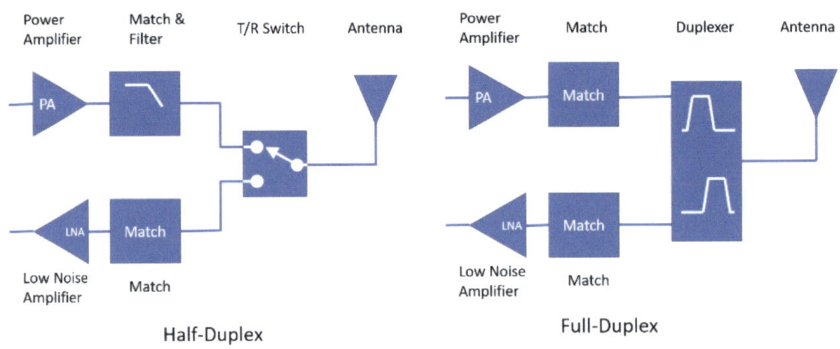

Figure 15.
Half-duplex and full-duplex front ends.

their own modules. The recent trend has been to take the LNA out of the transceiver chip and integrate it back with the RF front end in a module.

In the half-duplex radio, the filters can be relatively loose and made using standard passive components, so they tend to have less loss and can even be integrated with the power amplifier easily. Both the PA and LNA require matching circuits. For the PA the match is a power match and can often be combined with the trap filtering required. For the LNA a noise match is needed. The switch loss needs to be added to the filter loss on the transmit, but generally, the overall loss can be less than 2 dB. A receive channel filter is optional, and, in fact, modern, well-designed receivers do not need it, and so on the receive side, the front end can have less loss, be cheaper, and be relatively easy to integrate.

For a full-duplex radio, the duplexer needs to reject the transmit as much as possible on the receive side but also reject the transmit noise which lies in the receive band. The transmit noise in the receive band will be the limiting factor on the receiver performance. Duplexers are made using special processes—SAW, BAW, or FBAR—and add a significant cost to the front end. They tend to have a loss of 3 dB or more in the transmit path which means half the power put out by the PA is lost in the output network which directly translates to a need for a bigger battery.

The block diagram for a TDD radio front end is shown in **Figure 16**. Duplexing is done using time separation of transmit and receive, and the same frequency band is used for both. This means no duplexer is needed and the filtering can be handled by the same band filter for both transmit and receive. In some applications this filter can be omitted, but in applications where performance out of band is important (some of the TDD versions of the cellular standards and WiFi, for instance), it is often needed.

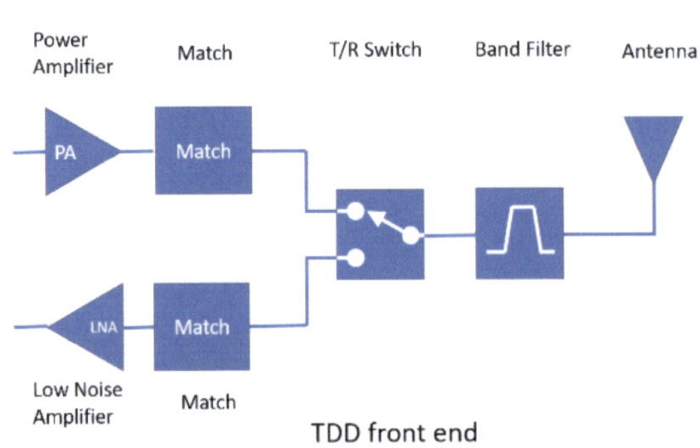

Figure 16.
TDD front end with band filtering.

4. Trade-offs in radio systems for communications

There are five choices that directly affect the amount of current needed in the RF front end for the transmit operation:

1. Frequency of the transmitter

2. Bandwidth of the signal

3. Number of radio bands supported

4. Type of modulation—linear or non-linear

5. Full-duplex or half-duplex

6. Output power

For the receiver, the list looks similar:

1. Frequency of the receiver

2. Bandwidth of the signal

3. Type of modulation—dynamic range required

4. Full-duplex or half-duplex

5. Presence of interfering signals

We will look at each individually.

4.1 Frequency of the transmitter and receiver

The frequency that the transmitter and receiver work at determines the current required. The higher the frequency, the more current are required as the transistors have less gain at higher frequencies. This immediately constrains the bandwidth of communication that can be handled as the bandwidth available increases with increase in frequency.

Another advantage of going low in frequency is that the signal will suffer less loss and can travel through obstacles better. If an IoT system is needed to operate throughout a building including traveling through walls, then it is better to operate at a lower frequency than at a higher one. This is readily seen in WiFi where the 2.4 GHz signal will be found to have more reach than the 5 GHz signal from a dual band router.

One disadvantage of using a low frequency is that the antenna size increases. If the system requires many antennas (as a MIMO system does), then this can be problematic.

4.2 Bandwidth of the signal

The bandwidth of the signal determines the noise level of the system as white noise is being integrated over the bandwidth. As the bandwidth increases, the noise level increases with it, so to maintain the same signal-to-noise ratio, the signal must be increased which increases the current. Also, at a given bias current, the gain-bandwidth product of transistors is fixed. As bandwidth increases, the bias current of transistors needs to increase to increase the gain-bandwidth product and maintain the gain at the same level.

4.3 Number of radio bands supported

Increasing the number of bands requires that some mechanism be found to stop all the circuitry for the radio bands that are not being used interfering with the transmission in the band that is being used. The usual method used to isolate bands is to incorporate a "band switch" in between the antenna and the output of each band. This switch is turned on when the band is to be used. However, when the power amplifiers for the radios are all on the same die, coupling increases, and the harmonics reaching the switch may be high enough that the switch does not have enough isolation. In the case where the switches are all on one die (the cheapest option as they can share circuitry), then the isolation can be further reduced by coupling.

The biggest issue with supporting a large number of bands is the necessity of providing filtering on these bands, especially in the case of full-duplex radios. As the number of bands and, therefore, filters increases, the cost increases. This is likely to lead to IoT solutions needing a separate RF front end for each geography covered so as to keep the number of bands manageable.

4.4 Type of modulation: linear or non-linear

The transmitter for non-linear modulation offers significant savings in power and complexity over a transmitter for linear modulation. A frequency synthesizer is all that is needed for most non-linear modulations, whereas a linear modulation would need the frequency synthesizer and a linear upconversion transmit. Finally, the power amplifier for a non-linear modulation can be a switching power amplifier; a linear modulation needs a linear power amplifier which is not as efficient.

Linear modulations offer the prospect of much higher bitrates, but this comes at the expense of power consumption. If you are trying to run at high data rates, then you will have to accept that your system will be power hungry.

4.5 Type of modulation: dynamic range required

On the receive side, the same type of receiver RF and analogue circuits are used for both linear and non-linear modulations, and so there is no difference seen between these types of modulation. However, the dynamic range required for the modulation has a large effect on the current requirements of the receiver. In general, a switch to higher orders of QAM means that a higher signal-to-noise ratio is required of the received signal than modulations like QPSK or GMSK, and so this means a wider dynamic range is required in the overall receiver.

4.6 Full-duplex or half-duplex

A full-duplex radio offers the prospect of higher bitrates on both the transmit and receive because both are working at the same time. Unfortunately, this means that more spectrum must be used to cover the transmit and the receive functions. A duplexer is necessary in this case, and this not only adds cost and complicates the front end design but also increases the loss in the transmit path. This increased loss leads to a lower efficiency for the transmitter and higher current draw for the same output power. On the receive side, the filtering must remove the large transmit signal, and this tends to lead to a larger loss for the desired signal as well. This loss must be compensated for with a lower noise figure in the receiver circuits, and this requires more current.

Half-duplex radio front ends are far simpler and have less loss. Most IoT data traffic is low data rate or bursty in nature, and this sort of traffic is well served by half-duplex radios. It should also be noted that the radio systems that use the ISM frequencies are half-duplex by necessity. As many IoT systems will make use of these radio systems, they are inherently half-duplex.

4.7 Transmit output power

Power is the product of current and voltage. Increasing the power requirements while keeping the voltage the same will require more current. For battery-powered devices, the voltage is fixed by the battery chemistry and by the reliability constraints of the integrated circuits being used, and so if more output power is required, then more current will need to be taken from the battery.

4.8 Presence of interfering signals

Large interfering signals on the receive side will affect the radio receiver by limiting the dynamic range of the receiver. If the signal is also large, then this may not be a problem. However if the desired signal is a lot smaller than the interfering signals, then the dynamic range of the receiver needs to be increased to receive both the interfering signal and the desired signal. In order to increase the dynamic range, it is necessary to increase the current in the circuits.

5. Overview of cellular radio systems

The most widely used and pervasive radio systems are cellular radio systems. Originally just an extension of the telephone service, cellular radio systems have grown into much more enabling data download and upload and giving access to the Internet virtually anywhere. For IoT applications that need mobility or access to the Internet from remote locations, a cellular radio will be the first consideration, and so we need to consider the alternatives. It should be noted that the need to pay royalties for IP included in the cellular specifications may make cellular a more expensive option than some of the other radios.

Cellular radios work on the principle of multiple transmitters and receivers separated into cells (hence the name), each controlling the communication with end users in the cell. The system works based on spatial separation: adjacent cells use a different set of frequencies, but cells sufficiently far away can re-use the same set of frequencies. Hand-off of mobile users is an important function of a cellular system, and a lot of resources are assigned to this function. A cellular network of cells each with its own base station is shown in **Figure 17**.

5.1 Introduction

Cellular radio systems are so-called because the coverage area is broken up into adjacent cells, each covered by a single wireless communication receiving and transmitting station called a base station. The original cellular radio systems were analogue and did not use digital modulation or any of the other techniques used in

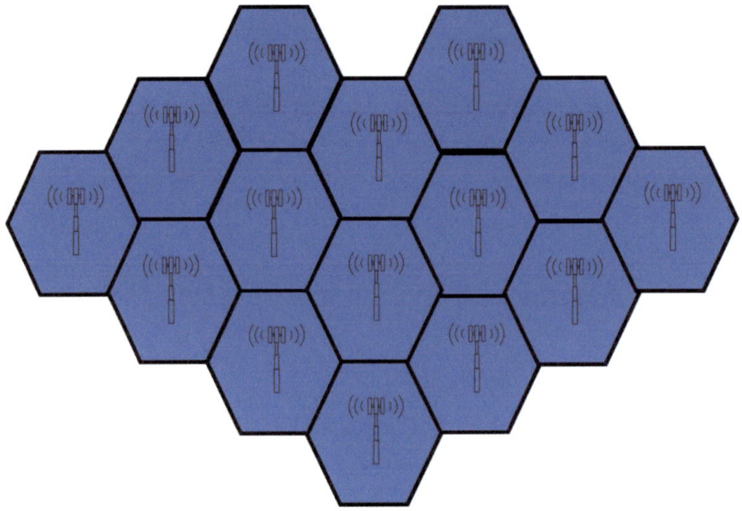

Figure 17.
A cellular radio layout with a base station at the centre of every cell serving that cell.

digital communications systems. We will not discuss those as they have been superseded and are of no relevance to IoT.

The first digital systems were the so-called 2G systems—analogue-based systems being 1G. There were multiple 2G systems deployed around the world—IS-136 (also known as TDMA) and IS-95 (generally called CDMA) in the USA and some other countries, PDC in Japan, and GSM throughout Europe and later most of the rest of the world.

The original 2G systems were circuit switched and served the voice market. Later, GPRS, a packet switched network, was added to the GSM system and allowed data communications to happen over the original GRPS network. Enhanced Data Rates for GSM Evolution (EDGE) was an upgrade of GPRS with a higher data rate. Some people used the names 2.5G to describe GPRS and 2.75G for EDGE.

High-speed data communications came with the 3G networks. All systems for 3G communications were based on CDMA technology. The GSM world migrated to Universal Mobile Telephone System (UMTS), while IS-95 CDMA networks migrated to CDMA2000. Further enhancements to the UMTS system— high-speed downlink packet access (HSDPA), high-speed uplink packet access (HSUPA), and high-speed packet access (HSPA) with further evolved versions adding a "+" (HSDPA+, HSUPA+, HSPA+)—allowed even higher data rates. The IS-95-based networks were able to upgrade to a new standard called evolution data optimized (EVDO). Following the naming convention adopted for higher-speed 2G networks, these enhancements were called 3.5G and 3.75G by some.

The GSM-based and IS-95 CDMA-based worlds came together with Long Term Evolution (LTE). For the first time, the world had a single cellular standard. This was named 4G by the marketing folks even though it did not meet the requirements for a next generation system as defined by the International Telecommunication Union (ITU) in its International Mobile Telecommunications Advanced (IMT Advanced) specification. In particular, while LTE is able to meet speeds of 100 MBps in the downlink and 50 MBps in the uplink, the IMT Advanced specification calls for 100 MBps to a fast moving vehicle and 1 GBps to a stationary device. The modulation in LTE changed to orthogonal frequency-division multiple access (OFDMA) on the downlink and single carrier frequency domain multiple access (SC-FDMA) on the uplink.

An enhanced version of LTE—LTE Advanced—does meet the requirements for a next generation system as defined by the International Telecommunication Union (ITU) in its International Mobile Telecommunications Advanced (IMT Advanced) specification.

The latest cellular standard, 5G, is being deployed now. With 5G, data rates have been increased up to a promise of 2 Gbps for some networks in ideal circumstances and latencies of 4 ms or less, with 1 ms being an oft-quoted target.

5.2 GSM/GPRS

The work to define a European cellular system started in 1983 under the Groupe Spécial Mobile (GSM) committee. This acronym later came to stand for Global System for Mobile Communications. The first systems were deployed in 1991. Over the years GSM and its packet radio version general packet radio service (GPRS) became the standard for cellular coverage. For IoT systems that need to work worldwide or in rural areas, GSM, or its IoT version extended coverage GSM IoT (EC-GSM-IoT), is still the system to use as it is the only one that is likely to have coverage.

GSM uses a combination of frequency division duplex (FDD), frequency division multiple access (FDMA), and time division multiple access (TDMA) to allow multiple users to use the system [4]. The transmit and receive frequency bands are separated (on the user side, the transmit is the lower-frequency band, and the receive is the higher-frequency band), with equal bandwidth dedicated to both. Each band is spit into channels of 200 kHz. Within a channel, TDMA is used to control access. Each channel is divided into frames of 4.615 ms each, and each frame is divided into eight time slots (or bursts) that are allocated between users. Users can use up to four of the eight slots in most systems, although there are higher-order specifications where more time slots are available, they tend not to be used. The maximum theoretically achievable bitrate is 114 kbps, but deployed systems do not get close to this limit.

The modulation used in GSM and GPRS is GMSK and is, therefore, a constant envelope modulation which lightens the requirements on the transmitter and power amplifier. Although there are classes of GSM that can be full-duplex, most implementations are half-duplex, and this removes the need for a duplexer. The combination of a constant envelope modulation and no requirement for a duplexer makes GSM transmitters very efficient.

The low bands for GSM allow for a transmission of up to 2 W (33 dBm) in the low band and 1 Watt (30 dBm) in the high band. The high output power with a timing advance specification in GSM that allows operation out to 35 km means that GSM is the ideal choice for operation in rural environments where, quite apart from the fact that it may be the only system available, the base stations will tend to be spread further apart.

It should also be noted that GSM has the capability to handle text messages in the form of short message service (SMS) which can be used to send small amounts of data. This capability could be useful for some IoT systems where only a small amount of data is typically sent. Each text message is 140 bytes although extensions allow more data to be sent by breaking that data up across multiple messages.

5.3 Edge

Enhanced Data Rates for GSM Evolution (EDGE) as the name implies is an extension of GSM and GPRS to further increase the maximum possible data rate [4].

The highest theoretical data rate is over 384 kbps, the threshold for a 3G system, but EDGE networks generally do not get near this number.

EDGE generally fits on top of the GSM system: it uses the same channel bandwidth as GSM (200 kHz) and similar filtering (Gaussian) and when the channel conditions are not good will use GMSK modulation. However, with the best conditions, it can use a new modulation: 8PSK with a $3\pi/4$ shift between symbols to avoid the zero crossing. The symbol for this new modulation is three bits long, and so, theoretically, three times the data can be sent over the channel. The 8PSK modulation has an amplitude component, and the strict adjacent channel requirements inherited from GSM mean that, unfortunately, the power amplifiers and transmit chain used in EDGE have relatively low efficiencies (higher teens percent than over 50% for GSM). This meant that there was a relatively large market for a hybrid solution with full EDGE capability on the downlink but only GPRS modulation using GSM components on the uplink to save on battery life. EDGE networks have mostly been superseded by 3G and LTE networks, and so it has little relevancy for IoT.

5.3.1 EC-GSM-IoT

Extended coverage GSM IoT (EC-GSM-IoT) is an extension to the GSM specification to allow it to work for IoT devices. The system was specified in such a way that the base station could be upgraded in software. As no hardware changes are possible, there are no changes to the modulation. Devices that support EC-GSM-IoT can be GMSK only or GMSK and 8PSK capable.

5.4 The UMTS: 3G system

The Universal Mobile Telecommunications System (UMTS) often known as Wideband-CDMA is the first true 3G system. The initial specification is known as R99 (Release 99) and is controlled by an organization known as 3rd Generation Partnership Project (3GPP) [5]. The GSM specification [4] was also rolled into 3GPP and is controlled by the 3GPP.

Several changes were introduced including the replacement of the TDMA used in GSM and EDGE with a code division multiple access (CDMA) to enable sharing of the medium. The bandwidth of the channel moved from 200 kHz to 5 MHz. The filtering was changed from a Gaussian filter to an approximation to a root raised cosine (RRC) improving the inter-symbol interference characteristics. The modulation used on the downlink is QPSK and OQPSK is used on the uplink. This means that the modulation has an amplitude content, but the adjacent channel requirements in UMTS are not as stringent as in EDGE, and therefore the PA needs less linearity and can be more efficient, with efficiencies over 40% on the uplink being common.

The system is full-duplex, meaning a duplexing filter (duplexer) is needed. This filter has more loss than the simple harmonic trap filtering that can be used in GSM because it is a half-duplex system and adds to the transmit losses of UMTS. Baseband processing in the system changed completely with the move to CDMA, and a lot more processing was added. Overall power consumption went up considerably with UMTS. Smaller geometries for the chips and new techniques have brought that power consumption down, but it is still higher than what can be achieved for a GSM system if that GSM system is optimized.

Later the air interface was changed and new modes added—high-speed downlink packet access (HSDPA), high-speed downlink packet access (HSUPA), and high-speed packet access (HSPA). There were many changes including changes to modulation available where the channel could support a higher data rate. In HSDPA 16 QAM was added to the downlink. Later a further enhancement was added in Release 7 of the 3GPP specification and was known as evolved high-speed packet access (HSPA+) which used up to 64 QAM modulation and promised theoretical speeds of 337 Mbps in the downlink and 34 Mbps in the uplink.

As a legacy system, there may be some IoT applications that make use of the 3G standards, but in many parts of the world, 3G has been superseded by Long Term Evolution (LTE) systems. For rural applications, GSM remains the only system with significant coverage throughout the developed and developing world.

5.5 LTE: popularly known as 4G

Long Term Evolution (LTE) is the standard that most networks, certainly in cities, are operating on. It is usually called 4G to differentiate it from 3G even though it does not meet the criteria for 4G as given by the ITU. LTE has theoretical data rates up to 300 Mbps on the downlink and 75 Mbps on the uplink [6]. The system is designed to operate out to 100 km with what is defined as acceptable performance and so could be used in a rural setting. The bandwidths used for the signal are flexible and vary between 1.4 and 20 MHz. LTE is available in frequency division duplex (FDD) and time division duplex flavors. With the addition of carrier aggregation, even wider bandwidths are effectively available at the expense of complexity in the specification, design, and testing.

The modulation was changed to orthogonal frequency division multiple access (OFDMA) for the downlink and single carrier frequency division multiple access for the uplink to limit the peak-to-average ratio (PAR) so that the power amplifier does not have to operate backed off by a lot and so conserving power. With special filters and power supply control, the power consumption of an LTE smartphone is reasonable in an urban context with base stations spaced relatively closely.

The change to OFDMA increased the spectral efficiency of LTE vs. UMTS by a factor of up to five times. As spectrum must be bought from the government for substantial fees, this is obviously attractive for cellular carriers. We should bear in mind, however, that spectral efficiency may not be factor in an IoT application.

The LTE channel is divided by time and frequency into units called resource blocks (RB). Each RB is 0.5 ms long and 180 kHz wide made up of 12 15 kHz sub-carriers. A user can be assigned a minimum of 2 RB in a 1 ms sub-slot. The more resource blocks a user is assigned, the higher the data rate is available to that user. The number of resource blocks available is dependent on the channel bandwidth and varies from 6 for a 1.4 MHz bandwidth up to 100 for a 20 MHz bandwidth. As can be seen, as each RB is 180 kHz, the 1.4 MHz bandwidth has 1.08 MHz of used bandwidth and 160 kHz of guard band on each side. For the 20 MHz case, the guard band would be 1 MHz on each side.

While the whole world is covered by four frequency bands for GSM (two in most of the world and two in the Americas), LTE requires a much larger number of bands to support worldwide operation. This requires a large number of filters and switches as well as power amplifiers and puts the cost up substantially.

5.5.1 LTE-M

Long Term Evolution Machine (LTE-M)-type communication and Narrow Band Internet of Things (NB-IoT) are low-power wide area network (LPWAN) protocols from the 3GPP as an extension of LTE that allows the cellular network to be used as an LPWAN. LTE-M was developed as a result of a realization that the LTE as it stood was not suitable for IoT and machine communications. The LTE-M targets are low device cost, long battery life, ability to support many devices in each cell, and deep coverage. LTE-M aims to achieve these objectives by allowing half-duplex operation, going to a single antenna only, adding a lower power class, supporting a lower data rate, and operating at a lower bandwidth.

There are two sorts of LTE-M defined in the 3GPP specifications—LTE-CAT-M1 and LTE-CAT-M2—but LTE-CAT-M1 is the most commonly deployed. LTE CAT-M1 uses a 1.4 MHz channel—the smallest available in the LTE specification. The system can support half-duplex and full-duplex operation. It is a single-antenna system (does not support receive diversity) which brings down the cost and power consumption.

5.5.2 NB-IoT

NB-IoT is a narrow band specification that fits into the LTE standard. NB-IoT can also be deployed in GSM networks and even in the guard bands at the edge of each frequency band in LTE. NB-IoT is aimed at high-density, low-bitrate support of IoT devices. Release 14 added support for mobility to NB-IoT (asset tracking is one of its largest use cases), but the degree of mobility is not a great as for the other cellular standards. Maximum supported data rates in the latest version of the specification are 127 kbps on the downlink and 159 kbps on the uplink; as always, real-world data rates are substantially less than this.

NB-IoT only supports half-duplex operation which makes the front end design simpler. It is a single-antenna system which also reduces complexity, cost, and power requirements. It uses a 200 kHz bandwidth which is how it is compatible with GSM. Within the 200 kHz bandwidth, the channel bandwidth is 180 kHz which fits with LTE and is the width of 1 RB. The 180 kHz of usable bandwidth in the downlink needs to be compatible with LTE and so uses OFDM with up to a maximum of 12 of the15 kHz sub-carriers. The uplink is also compatible with LTE and uses SC-FDMA with 15 and 3.75 kHz sub-carrier options.

The latency is specified at less than 10 seconds which is at least two orders of magnitude slower than other standards. For systems that need a low latency, this is going to be an issue.

5.6 5G

The latest addition to the cellular radio family is known, simply, as 5G. The 5G specification [7, 8] continues a trend from the earlier specifications of increasing the data rates over the link and also improving latency. Although improvements in latency (the delay over the network) were part of the aims of earlier standards, they became one of the main aims for 5G. The stated objectives for 5G are:

1. Data rate—5G is designed to deliver data rates ranging from 50 Mbps to 2 Gbps.

2. Latency—5G stated design target is 4 ms with 1 ms often being quoted as a target.

The data rate target is achievable in the lab but is more difficult to achieve in the field. To get close to data rates in the Gbps range, it is necessary for the transmitter and receiver to be relatively close.

With 5G a lot of new spectrum is being opened up. The radio interface specification, 5G NR (New Radio), defines two bands in the spectrum for 5G—FR1 (sometimes called “sub-6 GHz” even though it is specified up to 7.125 GHz) and

FR2 or millimeter wave (usually known as mmWave). The FR1 spectrum is now also broken into low-band and mid-band in most discussions. Low-band spectrum starts at 410 MHz and includes the current cellular spectrum up to the 2.4 GHz ISM band. Mid-band is the spectrum above the 2.4 GHz ISM band up to 6 GHz (which is the cut-off for the 5 GHz ISM spectral band) although the 3GPP is specified out to 7.125 GHz. Although networks are being rolled out in the low band, the mid-band is where the highest number of new networks is being introduced as these frequencies offer the possibility of getting towards the desired bit speeds. Networks at the very lowest frequencies often do not offer much performance advantage over advanced 4G networks at the same frequency.

The mmWave band is new frontier spectrum for cellular radios. Whereas the 3GPP specifications are specified for unwanted transmissions up to 12.75 GHz, the new mmWave spectrum has bands from 24.25 GHz up to 52.6 GHz. Operating at these high bands opens up the possibility to have 400 MHz bandwidths for signals as 400 MHz at a frequency of 25 GHz is only 1.6% of the band frequency whereas it is 100% of the band frequency at 400 MHz. Unfortunately, these high frequencies are much more difficult to work with. The signals at these frequencies are easily blocked, and the available transistor gain is much lower requiring either extra current to drive circuits or special transistors to get more gain at cost. Any system that needs to use the mmWave frequencies will need to have more base stations to cover the desired area.

The modulation schemes used in 5G are essentially the same as those used in LTE. OFDMA is used on the downlink, and SC-FDMA is used on the uplink. Sub-carriers of 15 kHz are used but now 30, 60, 120, and 240 kHz sub-carriers have also been added [9]. The sub-carriers are modulated using QPSK as the base modulation and 16 QAM, 64 QAM, and 256 QAM can be introduced as the link quality improves. As in LTE, the waveforms have a high PAR even on the uplink, and this leads to lower efficiency from the power amplifier.

Even in some of the low bands, channel bandwidths of 100 MHz are available, and this will add extra complications to the design of the transmit circuitry and power amplifiers as the circuits can experience “memory” effects. In a circuit experiencing memory effects, the circuit performance changes due to the amplitude of the signal but does not change back in time to process a change in amplitude. Wideband modulations put a larger stress on the circuit because it must react faster. Memory effects require that circuits not be driven at their limits which means that they have to be backed off and are less efficient.

Another objective defined in the yet to be released ITU-2020 Standard for 5G to be able to service 1,000,000 devices in a square kilometer. This is a 100× improvement on 4G systems. This has been referred to as “Massive IoT”. The 5G specifications as they stand are too power-hungry for most IoT applications which do not need the bitrate and latency advantages of 5G. The 3GPP is working on IoT specifications to be included in the 5G specification in an equivalent way to LTE-M and NB-IoT for LTE and EC-GSM-IoT for GSM. At the time of writing, these specifications are not available.

6. Overview of other radio systems

Outside the cellular world, there are a number of radio systems that are designed with IoT applications in mind. These radios work in the unlicensed industrial, scientific, and medical (ISM) bands of the radio spectrum. The use of these bands does not require a license, but they are still regulated bands, and any radio working in them must meet the regulatory requirements. There are a number of ISM frequency bands available, but the most commonly used are the 2.4 GHz band which is heavily used by radio communications devices and the 5.8 GHz band that is used for wireless local area networks (LAN).

Frequency spectrum is available at 902 to 928 MHz in Region 2 (the Americas)—known as 915 ISM—which is popular for IoT applications because it is less than 1 GHz, and so signals that travel better and transmit powers up to 1 W (30 dBm) are allowed. In Europe (part of Region 1), the main low band of cellular communications overlaps with this band, and so it cannot be used. In Europe there is a band at 868 MHz as part of the short-range device (SRD) spectrum, and that is used, but the output power of the transmitter is restricted to 25 mW (14 dBm), so it is less useful than the 915 ISM band.

Many of the alternatives to cellular systems use base stations to connect to nodes although they usually refer to these "base stations" as gateways or access points (in the case of wireless LAN). Unlike the cellular system, there is usually no handover protocol defined, and leaving one gateway requires a hard break and reconnection with the new gateway.

A network with a base station sitting at its centre and multiple sensors communicating through it is known as a star network. In some of the technologies outside the cellular world, it is possible to have another sort of network where nodes connect to each other and communicate with one another. Such a grid of nodes is known as a mesh. Star and mesh networks are shown in **Figure 18**.

It is, of course, possible to combine star and mesh networks. Such a network is shown in **Figure 19**.

The radio systems we will look at are Bluetooth Low Energy (BLE), Zigbee, 802.11ah, LoRaWAN, and Sigfox.

6.1 Bluetooth low energy

Bluetooth Low Energy (BLE) is an extension to the Bluetooth radio family controlled by the Bluetooth Radio Special Interest Group (SIG) specifically for IoT applications where long operation on battery is desired. Bluetooth Low Energy is not compatible with the previous versions of Bluetooth basic rate/enhanced data rate (BR/EDR). It is part of the Bluetooth 4.0 specification [10] and can be used alongside BR/EDR. The aim of BLE is to provide communications at the same range as BR/EDR at lower power and lower cost. It gives up voice capability and is purely data based in order to meet these aims.

BLE uses the same frequencies in the 2.4 GHz ISM as Bluetooth BR/EDR, but instead of 79 channels of 1 MHz each, it uses 40 channels of 2 MHz each. The modulation is Gaussian frequency shift keying (GFSK) which is a non-linear modulation which simplifies the design of the transmitter and uses less power than

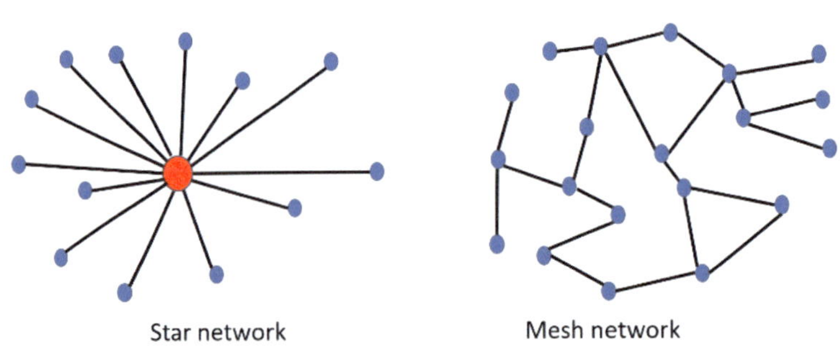

Figure 18.
Star network and mesh network.

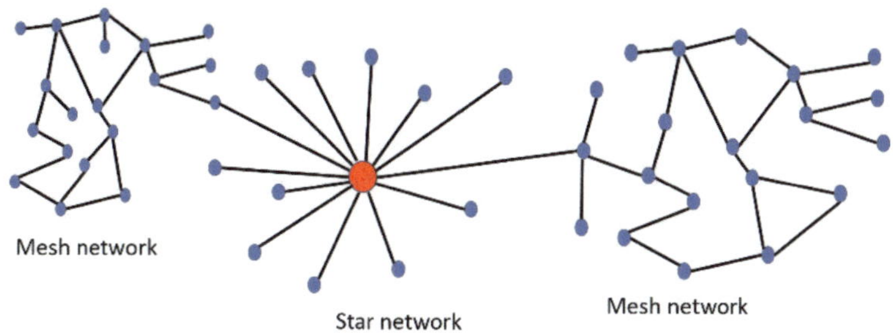

Figure 19.
Star network with mesh networks.

alternative linear modulations. In the Bluetooth 4.0 specification [10], the maximum output power was 10 mW (10 dBm). The maximum data rate is 1 Mbps.

One useful feature added with BLE is the concept of a mesh. BLE nodes can connect with their neighbors and pass a data transmission on to them. From one node it is possible to push a message across the whole mesh. This is another way of accessing an array of sensors rather than the central base station seen in cellular or gateways seen in other systems.

With the Bluetooth 5.0 specification [11], came some extensions to the BLE specification: higher power and higher data rate modes. The maximum power was increased tenfold to 100 mW (20 dBm) at a lower data rate of 500 or 125 kbps. This is expected to give a fourfold increase in range (power decreases with square of range from the transmitter, so a tenfold increase in power only sees a root-10 increase in distance). The maximum data rate was doubled to 2 Mbps but at a lower power.

BLE chipsets are designed to run off a standard coin cell battery and last up to 10 years. Although most BLE solutions will work for less than this you can still expect many years of operation.

There have been a number of cases of security vulnerabilities in Bluetooth being exposed. As ad hoc connections can be created, it is vulnerable to people pairing with devices which can, in the worst case, mean a loss of control. Also, one of the advantages of Bluetooth—the fact its communications can penetrate walls—is also a vulnerability as it can mean that a Bluetooth network can be accessed from outside a building.

Overall, Bluetooth, and in particular BLE, is an excellent radio protocol for IoT sensor nodes within a small geographic area providing the security issues are addressed. BLE offers no way to connect back to the Internet and hence cloud unlike cellular and some other systems; however it makes a good choice if the system needs to connect to multiple sensors through a local facility. In this case the system should also have computing at the edge capabilities to process data before sending the processed data into the cloud.

6.2 Zigbee

Zigbee is a networking protocol designed specifically for low-power and low-data rate devices. The Zigbee standard is developed by the Zigbee Alliance, but the physical (PHY—the lowest layer) and medium access control (MAC) layers are adopted from the Institute of Electrical and Electronic Engineers (IEEE) 802.15.4 specification [12]. This leaves only the network (NWK) and application (APL) layers in the Zigbee specification [13].

Zigbee is specified to work in the 868 MHz SDR, 915 MHz ISM, and 2.4 GHz ISM bands. In the 868 MHz band, one channel only is available, and BPSK modulation is used on that channel. Similarly, for the 915 MHz band, BPSK modulation is used, but there are 10 channels available. In the 2.4 GHz band, the modulation changes to OQPSK, and with the wider available bandwidth, 16 channels are available. ASK and OQPSK modulations are optionally available in the low bands for use in the case the 2.4 GHz channel is unavailable. Zigbee supports data rates in the 20–250 kbps range, so it is not suitable for many of the higher data rate applications but is well suited for many IoT sensor applications.

Output power in the 868 MHz band is limited to a maximum of 14 dBm which, with the BPSK PAR of about 2 dB, means the radio is limited to around a maximum of 12 dBm of output power. In the other bands, the regulatory limits are much high, and the highest power Zigbee radios can put out up to 20 dBm, but most transmit less. The line of sight range for Zigbee is quoted at 100 m maximum.

Two types of network can be set up with Zigbee devices: star and mesh. Combinations of the two types are possible. From this requirement, three different types of Zigbee device are needed: Zigbee coordinator, Zigbee router, and Zigbee end device. The Zigbee coordinator is the core of any Zigbee network. This is the node that will communicate with the outside world and control the whole network. Zigbee routers pass communications between each other but are not as capable as the Zigbee coordinator. Zigbee end devices will communicate only with either a Zigbee router device or the Zigbee coordinator and have minimal functionality. These will usually be the sensor nodes of the application. A Zigbee network is shown in **Figure 20**.

Zigbee is well designed for the purposes of connecting low data rate sensors in IoT applications at a local level. It is often criticized for not having enough security, so applications that use it may have to add their own data encryption on top which adds overhead and processing. It is not designed to connect back to the Internet and so to make the data available in the cloud.

6.3 802.11ah

The version of the wireless local area network (WLAN) specification designed to cover IoT usage cases is the IEEE 802.11ah specifications [14]. The other specifications in the 802.11 series (known as WiFi) are also useable in IoT applications but are primarily aimed at connectivity and are not optimized for IoT. The 802.11ah specification is also known as WiFi HaLow. While most of the wireless LAN specifications operate in the 2.4 and 5 GHz ISM bands (802.11ad, known as WiGig, operates at 60 GHz), 802.11ah will operate in the 868 and 915 MHz bands.

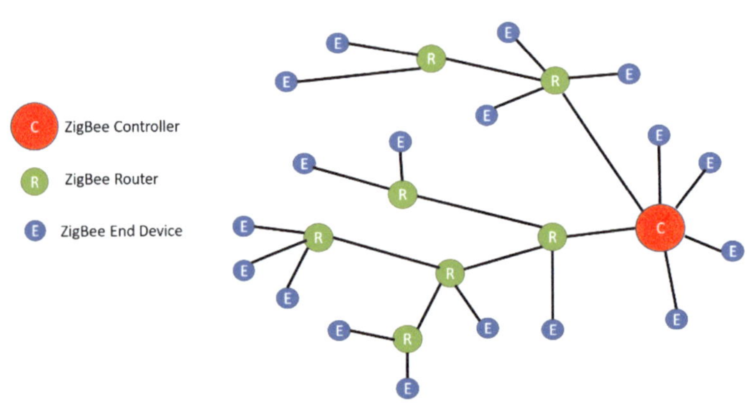

Figure 20.
Zigbee network illustrating the role of three device types.

WiFi HaLow is designed for longer haul communications than Bluetooth or Zigbee being capable of coverage up to 1 km. The system uses OFDM in 2 MHz channels, and the modulations used on the sub-carriers are all phase modulations going from BPSK up to 256 QAM. Available bitrates go from 150 kbps up to 234 Mbps.

Unfortunately, although WiFi HaLow offers some advantages—the ability to ramp the data rate over a wide range is a useful property to have—the standard has not been taken up by many companies, and no large company is producing chipsets to support it.

6.4 LoRaWAN

LoRaWAN is an low-power wide area network (LPWAN) networking technology built on top of the Long Range (LoRa) PHY layer protocol [15]. LoRa technology is available in the 433 and 868 MHz bands in Europe, the 915 MHz band in the USA and some other nations.

LoRa uses a unique modulation scheme: chirp spread spectrum [16]. The carrier is modulated much like an FSK signal, but in this case the frequency is either increased linearly or decreased linearly with a continuum of frequencies. Such a continuum of linearly increasing frequencies is called a chirp. Bandwidth is fixed, so the chirp can only move between a minimum and maximum frequency; however the rate at which it does that movement can vary. A chirp is shown in **Figure 21**.

There are three bandwidths available—125, 250, and 500 kHz—as well as six different slopes (known as spreading factors (SF)). For high data rates, the chirp will have a high slope and a correspondingly lower SF, and for low data rates, the chirp will have a low slope and high SF [17]. Start frequency determines the coding. On the receive side, the received symbol is multiplied with the inverse chip to extract the data. The higher the spreading factor, the longer the symbol that will appear in the receiver and the higher the likelihood of correct demodulation.

The LoRa modulation is a constant envelope modulation and as such lends itself to a compact transmitter and low power consumption. The nature of the modulation means that LoRa transmission can travel further for the same output power than many competing technologies. This makes LoRaWAN deployment attractive for rural and outdoor applications.

LoRaWAN is a networking technology built on top of the LoRa physical layer. The LoRaWAN network is a star network with individual nodes connecting back to a central gateway. This gateway is set up by the private company or individuals building the network, and this is one of the advantages of LoRaWAN—users build their own network and control their own data. Connection from this gateway to the Internet can be over Ethernet or fiber or can also be over the cellular network.

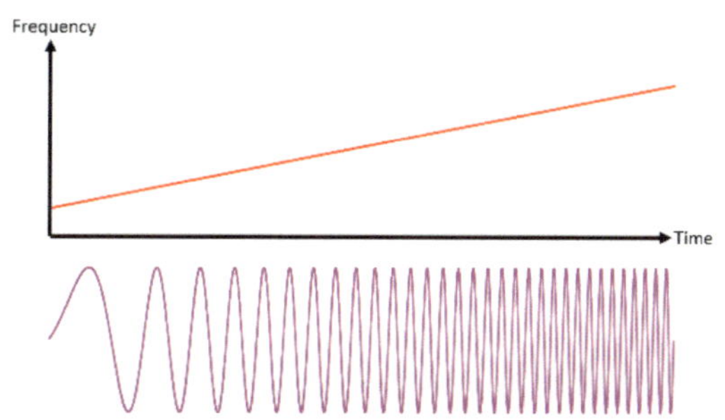

Figure 21.
Chirp signal frequency vs. time representation and time domain signal.

With LoRaWAN you are limited to 27 kbps data rate (there is an FSK option that can get up to 50 kbps). The other main limitation with LoRa technology is that it is only available on chips from Semtech Corporation. This limits the choice on performance and cost.

6.5 Sigfox

Sigfox is a French company that operates a network for IoT in the 868 MHz band in Europe and the 915 MHz band in the USA [18]. The Sigfox network is a star network with nodes communicating with base stations. Sigfox owns the base stations. The technology is designed for low data rate communications. The modulation is called Ultra Narrow Band (UNB) and makes use of a narrow bandwidth signal. The channels available are 200 kHz wide, but the bandwidth of the signal is only 100 Hz. With the narrow bandwidth signal, it is difficult to block the communication because any blocking signal needs to be right on top of the UNB signal. The carries are modulated using differential binary phase shift keying (DBPSK)—It is the value of the difference between a symbol and its previous symbol that is used for modulation not the symbol value itself and GFSK modulations. As the bandwidth of the communications is narrow, the data rate available is small—100–600 bits per second. Uplink messages are 12 bytes long, while downlink messages are 8 bytes long. A user is limited to 140 uplink messages a day and is only allowed to receive 4 downlink messages a day.

Sigfox is certainly something that should be considered if you have a low data rate application and do not want to communicate with the device that much—something that goes for a lot of sensor nodes. However, the message limits can be very restrictive.

6.6 Combining radio systems

As we have seen, there are many radio systems available and, those in this book are not an exhaustive list. However, every radio system has its advantages and disadvantages, and depending on the application, one radio system may be better than another.

Both BLE and Zigbee have the useful property of being able to support mesh networks. There are many instances where this may be useful: a network of sensors in a factory or a large office building or even a farm for instance. We could therefore envisage using Zigbee or BLE to network our sensors and bring the data back to a central place. At some point we will want to process that data. If the central location is in the offices of the factory, for instance, there may be no need to send the data any further. **Figure 22** shows a Zigbee network controlled from smartphone app using Bluetooth.

However, if the factory is part of a large network of factories, then the main office of the company may want to see the data. In this case we may want to make use of a cellular network to send the data back. In effect we have used the advantages of the two technologies—mesh technology for the local network and a secure, robust data link to send the results back. This would need some sort of local processing to put the data into a format that the main office can handle. **Figure 23** illustrates this with a Zigbee network connected through local processing to the cellular network.

We could send all the data from all the factories to the main office if we have a high-speed data link, but the head office may not want to process all that data. In this case, some form of computing at the edge is to process the data first and send a summary back (while storing all the data locally) to the main office. This also allows

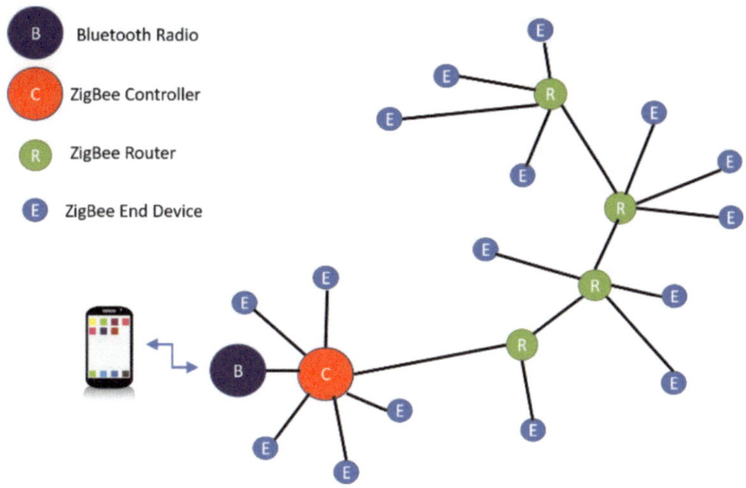

Figure 22.
A Zigbee network with Bluetooth link to smartphone control.

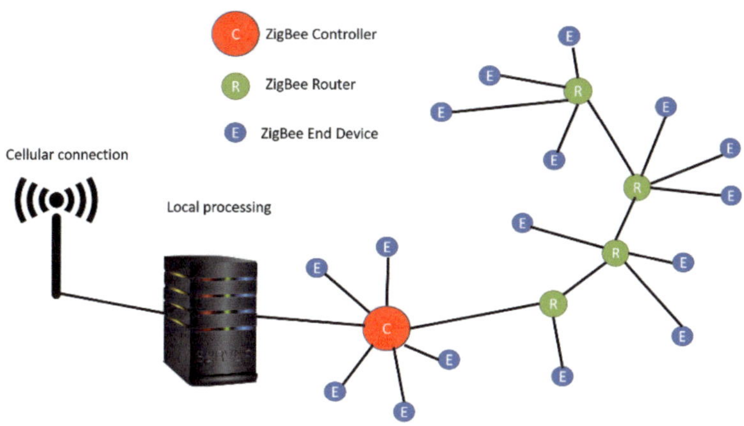

Figure 23.
Zigbee local network with local processing and cellular connection.

local management to catch problems quickly and start acting upon them before they get out of hand. If all decisions are taken at the main office based on data, they may miss problems because they have too many factories to check.

In the case of factories in remote areas with no access to high-speed data links, wired or not wired, it becomes necessary to compress the data sent back to the main office. In this case having local computing resources is a requirement. In fact, farms, oil facilities, mining facilities, highway rest stations, and many more places could all benefit by having networks of sensors linked up to a local processing facility with the data being sent back over another network.

7. Radio for sensing: introduction to radar sensors

We are familiar with radar in the context of ships and airplanes. These are large systems with large antennas. Recently advances in semiconductor technology have made it possible to integrate radars on chip. The initial application that started driving the development of radars was the reverse warning system for cars.

However, radars are now used for proximity detection in cars and are even being used in autonomous vehicle applications. With the availability of cheap integrated circuits from a number of manufacturers comes interest in using radars in a variety of applications.

Radar is divided into monostatic and bistatic types. Monostatic radars either use the same antenna for both receive and transmit or co-locate the antennas if they are separate. Bistatic radars separate the receive antenna from the transmit by a considerable distance. We will concentrate on monostatic radars here, not because bistatic radars are not interesting for IoT applications—they may well be—but because all the available integrated circuits are monostatic.

We can further divide monostatic radars into pulsed and continuous wave (CW) types. Most current radar chips are of a CW sub-type called frequency-modulated continuous wave (FMCW) which is the least complex yet powerful.

7.1 FMCW principles of operation

A CW radar transmits a continuous wave—a single frequency—which then bounces back off a target and is received back at the receiver. In the receiver it is mixed with the original transmitted tone, and this mixing brings it down to sit at 0 Hz (DC). Unfortunately this means we get no information about the distance of the object, but if the object moves, we see a shift in frequency received due to the Doppler effect, and this shift in frequency is seen as a shift away from DC in the downconverted signal. We can use this frequency shift to give us the velocity of the object. This is the principle of speed detection radar guns. CW radar is very cheap to build because it only requires a stable oscillator, downconversion circuitry, and some basic processing.

If we want to be able to detect distance—critical in applications like proximity warning systems—then we need to move to FMCW radar [19]. In FMCW radar, the transmit pulse is a frequency-modulated signal. Although it is possible to modulate the transmit signal with different signals, the most common and simplest is to use a sawtooth waveform or triangular pulse waveform. The signal in this case is known as a chirp. A chirp is a pure continuous wave which either linearly increases or decreases in frequency. Chirps, incidentally, are used in LoRa for communication. **Figure 24** shows the transmit and receive chirps represented as frequency vs. time and the relationship between them.

As can be seen, the transmit signal has changed frequency when the return pulse comes back. The difference in frequency may be small but it is measurable. Also as the transmit signal frequency slope is linear, the difference frequency, for a static

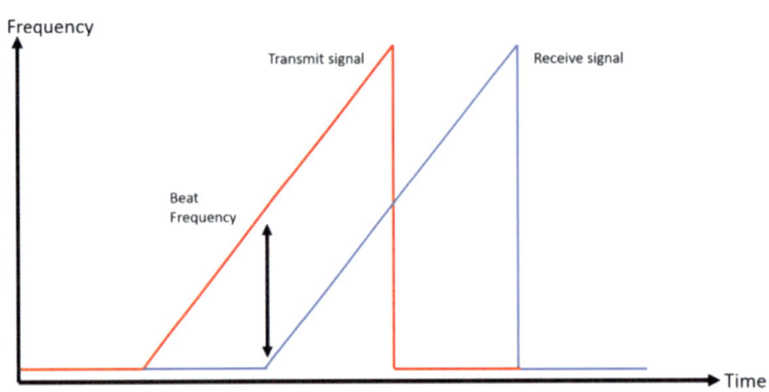

Figure 24.
Radar transmit and receive signals and their relationship.

object, will be a single frequency known as a beat frequency. From the beat frequency, we can calculate the distance to the object using the following equation:

$$f_{\text{beat}} = \frac{2 \times D}{c} \times \frac{df}{dt} \quad (15)$$

where f_{beat} is the beat frequency, D is the distance to target, c is the speed of light (3×10^8 m/s), and $\frac{df}{dt}$ is the slope of the chirp.

If the object is moving away or towards the antenna, then we will get Doppler shift in the frequency, and we will be able to detect that as well. We will not be able to detect it if an object is moving around a circle that is equidistant from the antenna. In effect we have a two-dimensional system—distance to the object and velocity. The velocity is given from the Doppler frequency using the following equation:

$$f_{\text{Doppler}} = \frac{2 \times v}{\lambda} \quad (16)$$

where f_{Doppler} is the Doppler frequency, v is the velocity, and λ is the wavelength at the radar frequency.

7.2 Moving to a 4D system

With a single antenna, we have distance to a target and velocity of the target relative to the antenna. If the antenna is idealized as an isotropic antenna (equal transmission in all directions), then we will have a sphere around the antenna at the distance to the target where the target could be. By adding a second antenna, we would have two spheres of potentially different size, and the target would lie somewhere on a circle at the intersection of the two antenna's spheres. Fortunately, real antennas are not isotropic and tend to be directional. In this case our spheres become distorted, and our circle becomes an arc. This is much more useful.

The antennas used with radar chips can be built into the package or are on the board. In either case they are patches of one sort or another which will transmit in much better away from the board or package than into it and behind it. By putting two antennas next to each other with at least half a wavelength distance at the radar frequency between them, we can determine where the target is on a plane stretching from left to right of the radar but not if it is above or below it. By adding a third antenna above the other two, we now know where, exactly, the target is in three dimensions relative to the antennas.

We already know that we can extract velocity towards or away from a single antenna but not the velocity of an object moving in a sphere around the antenna. However with more than one antenna, the object will always be moving relative to at least one of the antennas. So in the three-antenna configuration, the target will have velocity components that we can extract towards and at right angles to the path towards the antenna for each antenna. We can, therefore extract velocity information as well as direction of travel. This, in effect gives us four dimensions—the three special dimensions and velocity.

A common configuration of transmits in FMCW radar integrated circuits is two transmits and four receives. This configuration can be used to obtain a full 4D picture of the environment in front of the radar.

We can also extract other information about the target. In general, the return signal will also tell us how big the target is. This does, however, depend on the material the target is made of. Metal reflects radar radiation much better than, say, a human does or even a brick wall, and so a metal plate will appear relatively large. However, a human will reflect more than a dog, and so we are able to infer whether we have a human or an animal in front of the radar which may be useful in some remote monitoring applications.

7.3 FMCW signal processing

The processing required for FMCW radars is relatively simple. The digitized signal from the received signal is collected into a frame of a given number of samples, and this frame is fed into a discrete Fourier transform (DFT) usually implemented using an fast Fourier transform (FFT) algorithm. This gives us a spectrum of the signal where individual peaks will represent object and the frequency associated with that peak can be used to calculate the distance to the object. A second level of processing is then used to extract the velocity information. The spectra generated in the first stage are collected into another frame which now has two dimensions—time along one axis and frequency along the other. A DFT, again using an FFT algorithm, is run on this frame, and this gives us the velocity information at each frequency.

This processing needs to be performed for each antenna, and then the results need to be combined to give a full 4D picture of the world in front of the antenna. This signal processing is specialized but involves well-known algorithms and so is not particularly arduous to implement. FFT algorithm implementations are available in dedicated hardware blocks, and it is possible to build a hardware engine to perform the processing which would be significantly power-saving over a software implementation on a general-purpose processor. With dedicated hardware blocks to perform most of the processing required for the radar and a general-purpose processor to perform user programmed functions, we could build a flexible and high-performance imaging sensor unit which runs on relatively low power.

7.4 FMCW radar and video

Radar is a different sort of imaging sensor. It does not offer the sort of high-resolution images that are available from cameras, but it does have advantages over cameras in many applications. In general an FMCW radar installation will be of lower power and require less bandwidth from the communications systems than a video installation would. Even if the video images are processed locally, FMCW processing is a lot simpler than image processing for many applications and so requires smaller memories and computers and less power. Radar will also work in fog, rain, sleet, and smoke making useful in a wide range of applications. For applications that require identifying individuals or certain states (wearing a helmet or not, for instance), radar will not do the job, and image processing will be necessary. However, for some applications there are power constraints, a variety of atmospheric conditions may be encountered, or there is a privacy concern—monitoring of changing rooms or public toilets for instance—so for other applications, radar is a better fit. The next few sections will go over some of these applications.

7.5 Remote monitoring

FMCW radar is a good choice for monitoring applications. Video is used extensively in the security industry, but humans are notoriously bad at monitoring video feeds, and a system based on image processing and some sort of machine learning would be power-hungry. Generally, we are only interested in things that move—it is impossible to steal or damage something without getting close to it and physically moving it. Radar can tell you not only that something is moving but also how fast, in what direction, and even roughly how big the moving object is. It can also tell you how far away something is, so monitoring fluid level in a tank is also another good application of FMCW radar. Other non-security applications like automatic door opening (where the system can tell people are approaching the door rather than just passing) and people counting are also an extension of this monitoring.

7.6 Heart beat and breathing

For people that are close enough, FMCW is able not only to detect breathing but also heartbeat [20]. This is potentially extremely useful in many applications. An obvious example is disaster relief where radar can be used to detect people trapped in the rubble of a collapsed building. In home monitoring of the elderly is another potential application and one where radar is a better fit than cameras as it does not violate the privacy of the old person being monitored. There are numerous potential uses in the medical field including monitoring of sleep quality and looking for signs of sleep apnoea.

7.7 Gesture

Google has a project, Soli, that is developing solutions in using radar for gesture recognition [21]. Doppler radar can detect movement, and this movement can be deliberately coordinated to convey information—gestures. Gestures can be inferred by using machine learning (ML) to recognize which particular radar signature belongs to which particular gesture. Using this system gestures can be used to control equipment around us much like a television remote control.

7.8 Gait

As we have information about an object's position in all three cardinal dimensions and its velocity (with direction on the velocity), radar can be used to extract gait information. From the gait information, we will be able to identify a person from the way they walk. However, gait is also related to mental faculty. In fact, there is an ongoing medical research on using gait to give an early indication of neurodegenerative diseases like Alzheimer's disease and Parkinson's disease. In the future it may be possible to combine gait analysis with other analyses in a home monitoring application for the elderly based on FMCW radar devices.

8. IoT, radios, and edge computing

Cellular radio systems are set up with mobility in mind. Also, as the original systems were often voice, the transmit and receive systems were set up with equal bandwidth and data rate considerations. As data communications came to dominate, the systems that adjusted for the asymmetry in the data needs made the adjustment assuming the data needs would be higher on the downlink. For instance, streaming video needs a high bandwidth and high data rate on the downlink but not much on the uplink. This, as if happens, fits well with the requirements of mobile devices: mobile devices run off battery systems and so have limited power available. Keeping the upload data rate and modulation simple allows for a more efficient PA and hence saves power.

Many IoT systems will consist of fixed nodes sending data back to a fixed node with very little need (or possibly even ability) to receive data. In effect they need support for a higher data rate on the uplink than on the downlink. This is the opposite of the mobile client case, and so radio systems designed with a mobile client in mind may not be optimal for many IoT applications. It should also be obvious that the performance of the transmitter for an IoT sensor node is of much more importance than the performance of the receiver. In fact, in the simplest sensor nodes that continuously blast data out, there is not even a need to have a receiver.

8.1 IoT sensor systems and radio requirements

In an IoT system with the position of sensor nodes fixed, there is no need to support overheads in the radio system for mobile clients. This means overhead to support handover and multiple base stations are not needed. Fading as a result of the client device's movement will not be an issue; however changes in the environment around the radio may still contribute to fading although it should not be as deep as in a mobile application. This means it would be prudent to include interleaving.

The requirements for a wireless system for IoT depend on the usage case. For a system that sends back temperature data every few hours for instance, the wireless system used to send the data back would not need to handle a high data rate, but it would probably be desirable that it is robust and of low power. Also, the distance that needed to be covered would need to be considered, and so transmit power and transmit efficiency would be a factor in the system selection. With a low data rate requirement, the link robustness could be somewhat traded off as ARQ error correction would be possible. If the system needed to operate off a battery, the power dissipation of the wireless link would be a big factor if not the biggest factor in overall power dissipation. This would make the choice of wireless link critical.

A system sending back a continuous video feed would need to have a much higher data rate than a simple system returning a single sensor reading. It would need to be much more robust as it would not be able to make much use of ARQ without a significant bandwidth overhead. Battery operation is not practical unless the battery is large or there is some mechanism for battery recharging (solar, wind power, diesel generator).

It is possible to envisage a system which needs different requirements at different times. For instance, in a system to automate agriculture, we may use a combination of sensors strategically placed to measure temperature, moisture, sunlight, and other conditions. These sensors could be linked via a low-power radio system like Bluetooth or Zigbee back to a local control centre in the middle of the fields being monitored. The local control area processes the data from the sensors (an example of computing at the edge) and is linked back to a central control centre over a cellular link. The central control centre controls agricultural operations over a wide geographic area.

Normally, the local control centre would not need to send much data back—temperature, hours of sunshine during the day, light intensity, soil moisture, and so on. However, there may be circumstances where we need to inspect the location quickly—a sudden rise in temperature in one area indicating a fire and sensors in one area suddenly going down. As it is a remote location, we do not want to send someone out to check because it may be too late by the time the inspection team gets there. This is a realistic prospect in countries with a rapidly aging population where there is no manpower available to work in agriculture. In this case we would want to send a robot or even a drone over to inspect from the local control centre. If we must drive the robot or fly the drone from the central control centre, we will need a high bandwidth link to the local control centre and from there to the robot or drone. The solution is to put resources close to the drone or robot—people or computing. Having computing resources available at the edge to control the drones makes a lot of sense.

Even if the drone or robot is fully autonomous and the fields and local control centre are equipped with a 5G network so any video sent back is received at full definition, we still need a high-speed link to get that video data back to central control centre, so we can see what is going on. In this case we need to be able to ramp the communications speed up by several orders of magnitude which is not possible with today's radio technology. It may even be necessary to install a point-to-point wireless link if the only way decisions can be made is to get the data back. To get away from the need to have a high-speed link, it is necessary to move the computing resources close where they can monitor the video data. A system that can recognize a fire and send a message "there is a fire" is more useful than "we have a problem".

8.2 Bandwidth, bitrate, latency, and distance

As bandwidth increases, the bitrate that can be supported with the simplest modulations increases with it. If the bandwidth is fixed (as it is by regulatory requirements for wireless systems), then the only way to increase bitrate is to use deeper phase and amplitude modulations (16–64 QAM). A wide bandwidth system using a higher-order modulation will quickly run into physical limits: the higher modulation is pushing the acceptable noise level down as a larger signal-to-noise ratio (SNR) is needed, yet the wider the bandwidth means more noise, as noise is integrated over bandwidth. If you are transmitting at full power, then the only way to get more SNR is to move the transmitter closer to the receiver (or vice-versa).

If your application needs both high bitrate and needs to work over long distances, you need to reconsider your system. Is the sensor you are using appropriate? A lot of solutions use vision (a video camera) with AI software to perform their task. Video images consume a lot of bandwidth. Is there not some other sort of sensor that can do the job? Could it be done using radar, for instance? If the answer truly is no, then you will need to consider moving the computing closer to the camera. If you can do all or the bulk of the processing at the edge, what is the size of message that needs to be sent?

If the system has a high-bandwidth and a low-latency need, then the only choice will be 5G. The low-latency requirement will force you to move your computing to the edge because of the delay between the base station and the server. Although 5G promises 4 ms latency, the latency over the air is already twice that in current generation products, and the delay between the base station and the server can be 20 ms or more.

9. Conclusions

There are many radio systems available for IoT applications. Some may be suitable, and many will not.

Most IoT applications will not have access to either a wired power supply or a wired communications system. Systems built to work in these applications are, by default, going to need a wireless system to connect to the outside world. They are, by their nature, also going to be running off batteries, and so power consumption is a critical design specification for them. High data rate systems are going to consume more power than lower data rate systems. It is probably the case that moving the computing to the edge will remove the need for a high-speed link and may make it possible for a sensor node off a battery.

For wireless communications systems, a general rule is higher data rate requires more power. Also, the higher the bandwidth, the more difficult it is to set the system up: the position and angle of the antennas, what other devices are working around the system, and what are the obstacles in the way between the transmitter and receiver, amongst many reasons, all play into performance. Wherever there are high-speed data needs, the transmitter and receiver need to be close to each other. Adding computing closer to the edge can decrease the data requirements and make the implementation more practical.

One of the most useful and overlooked wireless systems is the SMS system that comes with all cellular systems. You can use SMS on 5G, 4G, 3G, and 2G systems throughout the world. It works on 2G systems in some of the more inaccessible locations. In order to make use of the SMS system, it is necessary in many instances to process the data first to compress it before sending. With a computing at the edge system, it becomes possible to make use of a truly useful system for getting data back.

Finally, if low latency is needed in a 5G system, the computing will need to be as close as possible to the base station. 5G promises 4 ms latency—we aren't there yet with more like 10 ms latency over the air only. If the data has to go back to the central servers to be processed, the latency would be more like 30 ms or more than 4 ms—hardly what 5G promised.

Author details

Malcolm H. Smith
AnalogueSmith (S) Pte. Ltd, Singapore

*Address all correspondence to: analoguesmith@gmail.com

References

[1] Rappaport TS. Wireless Communication Principles and Practice. 1st ed. Upper Saddle River, New Jersey: Prentice Hall; 1996. ISBN: 0-13-375536-3

[2] Bose JC. Detector for Electrical Disturbances, U.S. Patent 755,840, 1904

[3] Armstrong EH. Radio Signaling, U.S. Patent 1,941,068, 1933

[4] 3GPP. GSM/EDGE Radio Transmission and Reception, TS 45.005 Version 16.1.0; 2 April 2020; Third Generation Partnership Project (3GPP), 2020

[5] 3GPP. User Equipment (UE) Radio Transmission and Reception (FDD), TS 25.101 Version 16.1.0; 2 April 2020; Third Generation Partnership Project (3GPP), 2019

[6] 3GPP. Evolved Universal Terrestrial Radio Access (E-UTRA); User Equipment (UE) Radio Transmission and Reception, TS 36.101 Version 16.1.0; 8 April 2020; Third Generation Partnership Project (3GPP), 2020

[7] 3GPP. NR; User Equipment (UE) Radio Transmission and Reception; Part 1: Range 1 Standalone, TS 38.101 Version 16.1.0; 8 April 2020; Third Generation Partnership Project (3GPP), 2020

[8] 3GPP. NR; User Equipment (UE) Radio Transmission and Reception; Part 2: Range 1 Standalone, TS 38.101 Version 16.1.0; 9 April 2020; Third Generation Partnership Project (3GPP), 2020

[9] 5G Physical Layer Specifications—5G NR, Medium [Internet]. Available from: https://medium.com/5g-nr/5g-physical-layer-specifications-e025f8654981 [Accessed: 14 April 2020]

[10] Bluetooth SIG. Bluetooth Specification, Version 4.2; 2 December 2014; Bluetooth SIG. 2014

[11] Bluetooth SIG. Bluetooth Core Specification, Version 5.2; 31 December 2019; Bluetooth SIG. 2019

[12] IEEE. IEEE Standard for Low-Rate Wireless Networks, 802.15.4; 5 December 2015; Institute of Electrical and Electronic Engineers (IEEE). 2015

[13] Farahani S. Zigbee Wireless Networks and Transceivers. 1st ed. Burlington, MA: Newnes; 2008. ISBN: 9780750683937

[14] IEEE. IEEE Standard for Information technology—Telecommunications and information exchange between systems—Local and metropolitan area networks—Specific requirements—Part 11: Wireless LAN Medium Access Control (MAC) and Physical Layer (PHY) Specifications Amendment 2: Sub 1 GHz License Exempt Operation, 802.11ah; 7 December 2016; Institute of Electrical and Electronic Engineers (IEEE). 2016

[15] Technical Marketing Workgroup 1.0, LoRaWAN™ What is it? A Technical Overview of LoRa® and LoRaWAN™, LoRa Alliance [Internet], 2015. Available from: https://lora-alliance.org/resource-hub/what-lorawanr [Accessed: 14 April 2020]

[16] Mobilefish.com. LoRa/LoRaWAN tutorial 12: Modulation Types and Chirp Spread Spectrum; 2 October 2018; YouTube Video [Internet]. Available from: https://www.youtube.com/watch?v=lg0eZWZFKiE [Accessed: 14 April 2020]

[17] Mobilefish.com. LoRa/LoRaWAN tutorial 13: Symbol, Spreading Factor and Chip; 4 October 2018; YouTube Video [Internet]. Available from:

https://www.youtube.com/watch?v=0FCrN-u-Vpw [Accessed: 14 April 2020]

[18] Sigfox Homepage [Internet]. Available from: https://www.sigfox.com/en [Accessed: 14 April 2020]

[19] Autonomous Stuff. Introduction to Mmwave Sensing: FMCW Radars; 14 April 2018; YouTube Video [Internet]. Available from: https://www.youtube.com/watch?v=8cHACNNDWD8 [Accessed: 14 April 2020]

[20] Ahmad A, Roh JC, Wang D, Dubey A. Vital signs monitoring of multiple people using a FMCW millimeter-wave sensor. In: 2018 IEEE Radar Conference (RadarConf18); Oklahoma City, OK. 2018. pp. 1450-1455

[21] Google ATAP. Welcome to Project Soli; 30 May 2015; YouTube Video [Internet]. Available from: https://www.youtube.com/watch?v=0QNiZfSsPc0&t=3s [Accessed: 14 April 2020]

Chapter 6

Energy Harvesting Technology for IoT Edge Applications

Amandeep Sharma and Pawandeep Sharma

Abstract

The integration of energy harvesting technologies with Internet of things (IoTs) leads to the automation of building and homes. The IoT edge devices, which include end user equipment connected to the networks and interact with other networks and devices, may be located in remote locations where the main power is not available or battery replacement is not feasible. The energy harvesting technologies can reduce or eliminate the need of batteries for edge devices by using super capacitors or rechargeable batteries to recharge them in the field. The proposed chapter provides a brief discussion about possible energy harvesting technologies and their potential power densities and techniques to minimize power requirements of edge devices, so that energy harvesting solutions will be sufficient to meet the power requirements.

Keywords: energy harvesting, edge devices, IoTs, standards

1. Introduction

The technological advancement in energy-efficient low-power hardware vanquished the need of AC current for IoT-based embedded systems, making them suitable for remote applications. The number of new applications including weather data estimation and real-time parameter monitoring is gaining ground with the evolution of different environmental sensors and telemetry applications. Energy harvesting technologies together with low-power platforms and energy-efficient storage technologies allow edge devices, IoTs and embedded systems to work in remote areas.

A report published on energy efficiency of the Internet of things (IoTs) focuses attention on the utilization of edge devices in various applications, and most of these applications are battery driven, which are having limited lifespan (**Figure 1**). The objective of the proposed chapter is to highlight the potential of energy harvesting technologies that can reduce the dependency on fixed charge batteries and lead to interrupted device operation. The chapter covers the discussion on basic elements of an energy harvesting system, enumerates possible energy resources and quantifies the potential of different harvesting technologies including photovoltaic cells, piezoelectric materials, thermoelectric and electromagnetic generators and electrostatic motors.

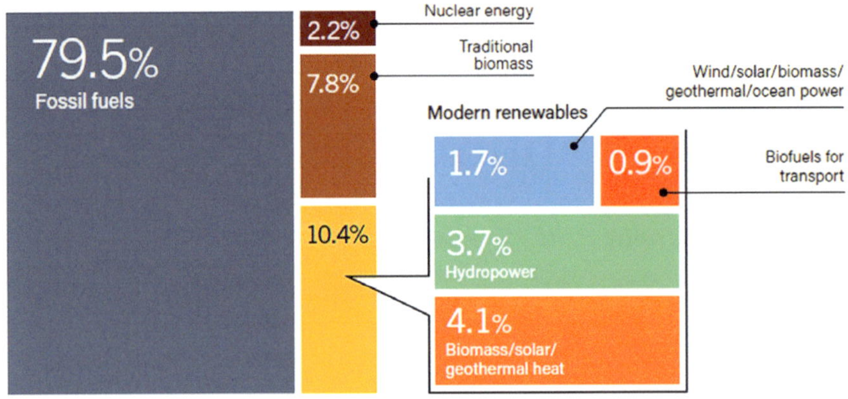

Figure 1.
Share of renewable energy resources in total energy consumption, 2016 [1].

2. IoT and IoT edge devices

Internet of things abbreviated as IoT leads to ambient intelligence that is based on the connectivity of people and things and utilizes the adaptive and sensitive electronic environment to address the need of the things around. Diab et al. [2] summarized six essential building blocks of IoT-based system, as shown in **Figure 2**.

Deployment of IoT-based systems in remote locations demands unattended operation over long time spans. In achieving this target, constant power supply and energy efficiency are the key challenges. Energy harvesting allows on-site charging of the storage devices and thus leads to an uninterrupted sensor node operation. In addition to it, edge devices control the flow of information between two network boundaries. Basically, the edge devices are utilized by service providers and act as entry and end points for a network. Their primary functions include processing, filtering, translation and storage of data and transmission and routing of data within the network [3].

In order to attain intelligence systems with advance features and to gain more computing power, IoT-based applications make use of edge devices. This process of using logical physical locations and decentralized processes is called edge computing [4].

Edge router is the most common type of edge devices and acts as the gateway between the different networks [5]. The primary function of the edge router is to connect a wide area network or a campus network with the Internet. Firewalls are the category of edge devices which are located on the margins of network and perform filtering of processed data during transmission between external and internal networks [6]. Further, different types of sensors/actuators and other types of end terminals also act as edge devices. The figure given below encapsulates the different types of end devices and their role in the network (**Figure 3**).

If the number of IoT devices is connected in a system, edge devices provide distributed operation among them using dynamic host configuration protocol and domain name systems [7].

2.1 Importance of edge technology

The conventional cloud computing network being a centralized network collects the data at the outermost layer and sends it to the server for further procedure. Limited hardware capabilities of the devices especially near the network edge

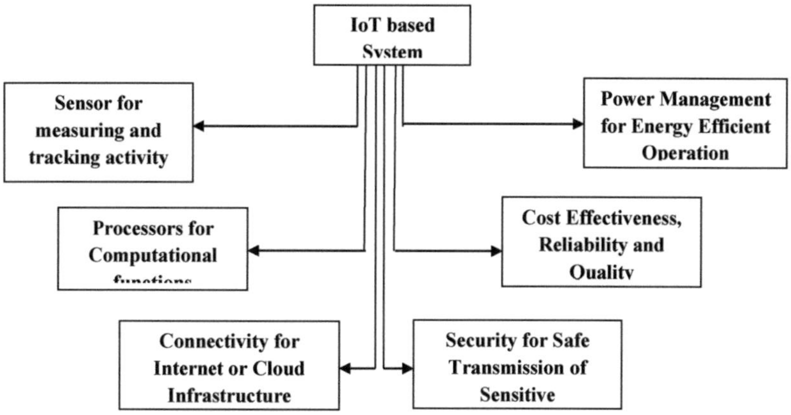

Figure 2.
Essential building blocks of an IoT-based system.

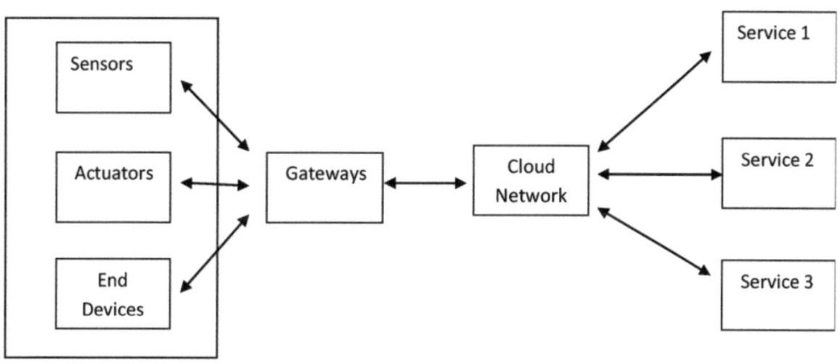

Figure 3.
Edge devices in IoT-based system.

are the main reasons behind this setup. These devices have limited functionality, limited power capabilities and limited storage capacity that restrict them to process or analyse the gathered data. With the advancement of miniaturization techniques in fabrication, present day IoT devices have capabilities of handling large data. This feature leads to optimized network operations by collecting data at edge terminals and relocating advance processing functions in real time [8].

Thus edge computing ensures processing of data where it is created rather than routing the data to data centres for processing. This feature leads to improved response time to milliseconds and optimum usage of network resources.

2.2 Advantages of enabling edge computing for IoTs

- **Reduced data exposure and reduced network load:** Real-time data handling by the edge devices with advance processing and storage capabilities prevent the data to route through the whole network and enhance the offline capabilities by making the apps independent from uninterrupted network connection.

- **Reduced delay:** IoT edge computing devices optimize the network performance by reducing delay or latency parameter. By storing and processing the data in edge data processing units, real-time computing is possible instead of communicating with the cloud server for each interaction.

- **Secure communication:** Since edge computing architecture deals with distributive nature of collecting, processing and storing the data among the large range of data centres and edge devices, it is not easy for an intruder to demolish the privacy and security of the network [9].

2.3 Industry benefits

- Efficient massive data processing
- Local data handling ensures security of sensitive data
- Quick response time with smart devices and applications

Cloud-based deployments of edge devices and data exchange features between edge and cloud is the next generation build out for 5G communication by telecommunication sector. High-speed network response with very low latency factor is the target where data compilation is carried out at the edge devices and reports are sent to the central cloud for storage. This feature eliminates the unnecessary data movement over the network [10].

3. Energy harvesting technology

Energy harvesting is a process through which energy is derived from external resources and captured energy is converted into electrical energy through energy harvesting device. The various resources are tabulated in **Table 1** that can be used for conversion.

Basically, an energy harvesting system comprised of three main components including application specific transducer, an interface circuit with or without storage device and receiver. The transducer or energy harvesting unit harvests the energy from the ambient sources and converts them in electrical form. The function of the interface unit is to extract maximum amount of energy from the harvesting unit and make the energy level compatible with the specific receiver or load. This has been accomplished with different power management approaches including voltage regulation or rectification [20], etc. The receiver may include different sensors, transducers or any other electronic circuit. The presence of storage unit avoids the start-up problem and energy depletion state in case of large interval in harvesting cycles. **Figure 4** shows the basic block diagram of an energy harvesting system:

S. No.	Form of energy	Source
1	Light energy	Solar energy from sun (outdoor/indoor) [11, 12]
2	Kinetic energy	Vibration, rotation, motion [13–16]
3	Thermal energy	Human body, industry [17, 18]
4	Atmospheric energy	Pressure, gravity
5	Radio frequency	Antennas, radio frequency spectrum [19]
6	Biological/chemical energy	Diffusion, radioisotopes
7	Hydro energy	Kinetic energy from water

Table 1.
List of various energy resources and their source.

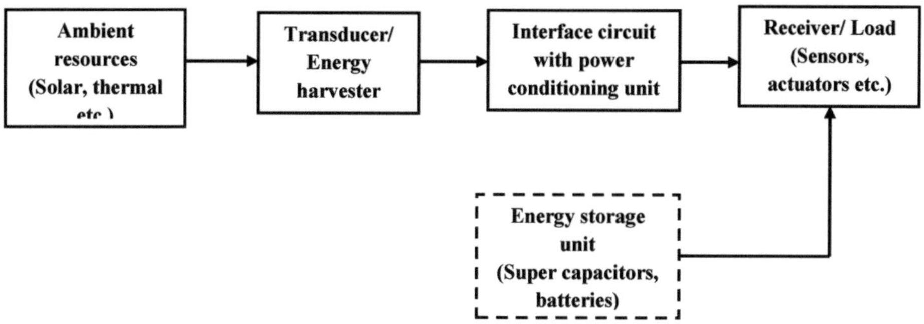

Figure 4.
Basic building blocks of an energy harvesting system.

4. Requirement for energy harvesting in IoT

Energy harvesting is a promising solution to power IoTs especially when they are installed in inaccessible areas and regular battery maintenance is not possible. Energy harvesting approach extends the life cycle of the device and eliminates the constraint of fixed charge batteries as an energy source. Some key factors are enlisted below that highlight the requirements for energy harvesting technology in IoT applications.

S. No.	Energy harvesting technology	Potential production	Industrialization
1	Thermoelectric	10 μW–1 kW	Widespread production
2	Photovoltaic	1 μW–1 MW	
3	Electrodynamic	0.1 μW–1 MW	
4	Piezoelectric	10 μW–100 W	
5	Capacitative electrets movement harvesting	0.1 μW–1 mW	Research
6	Pyroelectric		
7	Capacitive without electrets	0.1 μW–1 MW	
8	Triboelectric		
7	Radio frequency waves	0.1 μW–1 mW	Limited production
8	Magnetostrictive	10 μW–1 MW	Major trials

Table 2.
Different energy harvesting technologies and their potential production [21].

S. No.	Electronic module	Power range
1	Watch/calculator	1 μW
2	RFID tag	10 μW
3	Sensors/remotes	100 μW
4	Wireless sensors/hearing aid	1 mW
5	Bluetooth transceiver	10 mW
6	Global positioning system(GPS)	100 mW

Table 3.
Overview of power consumption by different IoT modules and sensors [22–24].

4.1 Power

The energy harvester should generate power at least of the order of milliwatts to sustain in IoT domain applications. **Table 2** shows the potential production of different energy harvesting technologies.

In conjunction to the survey given in **Table 2**, an overview of power consumption of different electronic modules has been depicted in **Table 3**. The survey reflects that the operating range of different IoT devices and sensors is in between 0.1 μW and 1 W, which can easily be handled by energy harvesting devices. Since ambient resources are stochastic in nature, energy demand and supply may not be time synchronized; the presence of backup storage devices and effective power management electronics is essential to deliver power from harvester to IoT devices in time.

4.2 Size scaling

With the advancement in integrated circuit technology, the size of the IoT devices is not an issue as number of features can be integrated on a single chip.

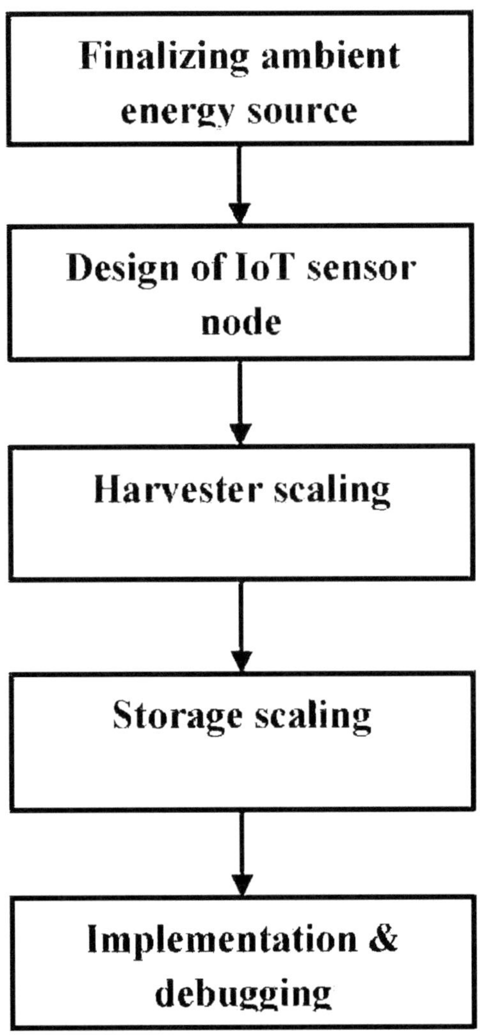

Figure 5.
Development procedure of rechargeable sensors.

The battery used in conventional module designs generally has a life cycle of 1 year and is the key factor in overall weight and size of the module. As an alternative to the fixed charge batteries, the size of the energy harvesting unit is application specific and should not be greater than the previous energy storage. The scalability of the energy harvesting unit with the size of IoT module should be ideal.

4.3 Cost factor

Conventional battery is an economical product because of the mass production, which leads to cost-sensitive production of battery-driven IoT devices. On the other hand, integrating energy harvesting technique into the module will increase the cost. This cost will include the component cost together with the redevelopment cost of the device because implanting an energy harvesting device on the top of the conventional module is not a practical solution and whole internal design gets modified. **Figure 5** shows the development process of rechargeable sensors and validates the above statement.

5. Constraints and potentials of energy harvesting technique

5.1 Constraints

The constraints in practical implementation of energy harvesting techniques in IoT domain are as follows:

1. High cost as compared to the conventional batteries is the most important constraint to be considered.

2. The second barrier is the size of the harvesting module, which increases as per the energy demands.

3. Energy storage unit is essential for uninterrupted sensor node operation.

5.2 Potentials

Although the additional cost is a barrier for mass production of harvesting modules and power requirements of some technologies may not be in the range of harvesting technologies, there are high possibilities in wide spread adaption of energy harvesting technology. Few of them are enlisted below:

1. The estimated life cycle of an energy harvesting module is above 5 years. In a survey, it has been observed that some energy harvesting units are working smoothly from the last 15 years without hardware degradation [25]. Thus, regular hardware servicing is not required in inaccessible locations.

2. The evolution of advance low-power electronic hardware together with the cloud computing for data processing further reduces the energy consumption in electronics and increases the feasibility of energy harvesting.

3. A high factor of energy saving can be in building automation where copper wires, materials, installation and maintenance cost can be reduced by energy harvesting approach.

4. With the growth of IoT industry, more energy harvesting modules will be implemented that will in turn reduce the cost.

6. Energy harvesting standards

Interoperability between different end systems is an important criterion for the successful adaption of energy harvesting techniques in different applications. To achieve this target, there are three standard policies, which are depicted in **Figure 6**. The EnOcean Alliance follows the standardization of communication profiles (EnOcean Equipment Profiles) to ensure that the entire product range including rechargeable sensors, wireless switches and controls can communicate with each other. The EnOcean wireless standard is geared to wireless sensors and wireless sensor networks with ultra-low-power consumption and also includes sensor networks with energy harvesting technology.

The EnOcean wireless standard became the standard ISO/IEC 14543-3-10:2012 in 2012. The standard is applicable to Information technology, Home electronic system, wireless short packet protocol and optimized for energy harvesting, its architecture and lower layer protocols. The proposed protocol supports energy harvested products for IoT sensors and switches designed without wires and batteries. The standard allows low-power consumption of sensors and switches by transmitting multiple short transmissions and appropriate frequency bands with adequate signal propagation and minimum interference.

6.1 Socio-economic applications

In this section socio-economic application of energy harvesting technology has been discussed, and their market share has been depicted in **Figure 7**.

- **Energy harvesting power sensors**

  Renewal of battery-operated power sensors with energy harvesting-based rechargeable wireless sensors is the prime factor in the growth of energy harvesting market. Rechargeable sensors are becoming the first choice for the deployment

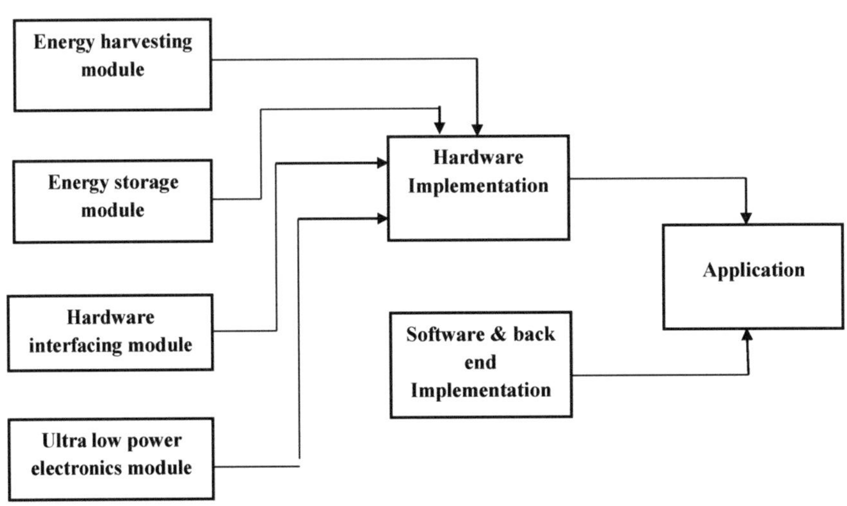

Figure 6.
Implementation of energy harvesting platform.

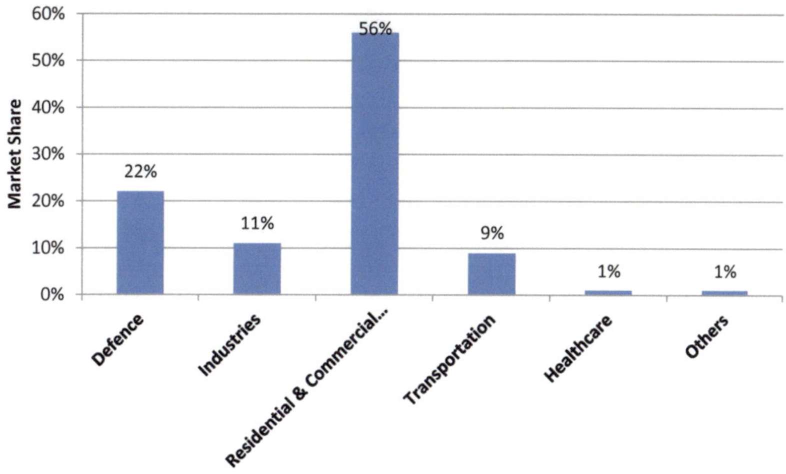

Figure 7.
Energy harvesting market share under different sectors.

in remote or inaccessible locations where it is not possible to replace the batteries frequently especially in offshore oil, gas systems or sensitive military areas [26]. Further, autonomous or rechargeable sensors are environment friendly as they do not contain any harmful metal or chemical for the environment.

- **Benefits for primary industry**

The energy harvesting market in primary industry sector was sized at EUR 130 million in 2014 and predicted to reach EUR 380 at the end of 2020 [27]. To enhance the reliability and availability of industrial processes, energy harvesting devices are employed in primary industry sector.

- **Benefits for defence industry**

Energy harvesting technology has been increasingly used in defence sector as it leads to the way where heavy battery packs are replaced by autonomous or rechargeable devices. Solar photovoltaic-based autonomous aerial unmanned vehicles called Drones are the great achievement for military applications. According to the latest survey, energy harvesting-based defence applications target EUR 730 million by 2020 [27] (**Table 4**).

- **Benefits for transport industry**

Different transportation media offer various forms of energy that can be harvested by different means. For instance, solar panels installed on the road collect the solar energy and convert it into electrical form through walking or driving. Similarly, vibration energy generated through various transportation means can be utilized for energy harvesting. It is estimated that transportation sector can target EUR 310 million by 2020 [27] (**Table 5**).

- **Benefits for residential and commercial sites**

Energy harvesting will be an essential feature for the upcoming trend of home automation and smart housing. This is because of the reason that energy

	Application	Edge devices	Energy source
Smart industry	Remote control of machines	Machines	Main power supply
	Real-time monitoring of machine wear	Appropriate sensors	Main power supply
	Diagnostics of machines	Appropriate sensors	Main power supply
	Surveillance of maintenance conditions	Appropriate sensors	Main power supply; energy harvested source
	Inventory management and asset tracking	RFID tag	Passive energy source; energy harvested source
	Smart pipeline management	Dedicated sensors, e.g. thermal, pressure, humidity, etc.	Energy harvested source

Table 4.
Application and energy sources of edge devices for smart industries.

Application area	Application	Edge devices	Energy source
Smart transportation	Smart roads	Energy collector roads	Pressure- or vibration-based energy harvesting technology
		Smart sensor in roads	Vibration, RF or thermal-based energy harvesting technology
		Smart lightening system for roads	Main power supply; solar- or vibration-based energy harvesting system
	Real-time monitoring of traffic congestion	GPS system in cars	On board or rechargeable battery
	Car-to-car communication system	Inbuilt devices in car	On board or rechargeable battery
	Car-to-infrastructure communication system	Inbuilt devices in car	On board or rechargeable battery; solar- or vibration-based energy harvesting system

Table 5.
Application and energy source of edge devices for smart transportation.

harvesting technology will reduce the large-size batteries for mobile devices. Rapid growth of IoT devices is also possible with energy harvesting as it facilitates rechargeable power supply with low installation and maintenance cost. The market value of this sector is expected to reach EUR 1750 million at the end of 2020 [27] (**Table 6**).

7. Communication protocols used in IoT systems

Various communication protocols with their different positive and negative attributes have been used to communicate with the network through IoT devices. **Table 7** enlists different communication protocols along with their properties.

8. Summary

Different levels of technical maturity are available with different types of ambient energy resources and corresponding energy harvesting techniques.

Target area	Application	Edge device	Energy source
Smart home	Home automation	Actuators	Main power supply
		Camera	Main power supply
		Gateway	Main power supply
		Smart sensors including humidity sensor, temperature sensor, fire sensor and door sensor	Energy harvesting source of light or thermal energy
		Sensors connected with mains	Main power supply
	Smart lightening system	LED bulbs	Main power supply
		Hub; gateway	Main power supply
		Smart light switches	Vibration- or pressure-based energy harvesting source
Smart workplace	Workspace automation	Main server	Main power supply
		Actuators	Main power supply
		Sensors	Energy harvesting source of light or thermal energy
		Gateways	Main power supply
	Control system	Reporting time, access control	Main power supply
		Badge	Passive source of energy
		Gas sensor, motion sensor, smoke sensor	Energy harvesting source of light or thermal energy
		Window/door sensor	Energy harvesting source of light or vibration energy

Table 6.
Application and energy source of edge devices for smart home and working area.

Communication protocol	Power	Data rate	Distance covered
ZigBee	50 mW	250 kb/s; for medium data rate applications	25 m; for medium distance communication
Wi-Fi	Greater than 500 mW	Greater than 100 Mb/s; for high data rate applications	10 m; for low distance communication
LoRa	100 mW	1 kb/s; for low data rate applications	500 m; for high distance communication
Bluetooth	10 mW	1 Mb/s; for high data rate applications	10 m; for low distance communication

Table 7.
Different communication protocols with their properties.

The magnitude of available power from different sources ranges from microwatts to milliwatts. This range is suitable for a variety of IoT devices and related applications. Some practical implementations of energy harvesters are surveyed in the literature that offers power to drive IoT devices. But there are various technical constraints that limit the worldwide implementation of energy harvesting technology. These issues need to be addressed to satisfy the future demands of the world.

Author details

Amandeep Sharma[1*] and Pawandeep Sharma[2]

1 Electronics and Communication Engineering, Chandigarh University, Punjab, India

2 University Institute of Computing, Chandigarh University, Punjab, India

*Address all correspondence to: amandeep.ece@cumail.in

References

[1] Gondal IA. Offshore renewable energy resources and their potential in a green hydrogen supply chain through power-to-gas. Sustainable Energy & Fuels. 2019;**3**(6):1468-1489

[2] Diab A, Mitschele-Thiel A. Self-organization activities in LTE-advanced networks. In: Handbook of Research on Progressive Trends in Wireless Communications and Networking. Pennsylvania, US: IGI Global; 2014. pp. 67-98

[3] Hassan N, Gillani S, Ahmed E, Yaqoob I, Imran M. The role of edge computing in internet of things. IEEE Communications Magazine. 2018;**56**(11):110-115

[4] Chen B, Wan J, Celesti A, Li D, Abbas H, Zhang Q. Edge computing in IoT-based manufacturing. IEEE Communications Magazine. 2018;**56**(9):103-109

[5] Jutila M. An adaptive edge router enabling internet of things. IEEE Internet of Things Journal. 2016;**3**(6):1061-1069

[6] Kubler S, Främling K, Buda A. A standardized approach to deal with firewall and mobility policies in the IoT. Pervasive and Mobile Computing. 2015;**20**:100-114

[7] Maheshwari N, Dagale H. Secure communication and firewall architecture for IoT applications. In: 2018 10th International Conference on Communication Systems & Networks (COMSNETS). IEEE; 2018. pp. 328-335

[8] Li H, Ota K, Dong M. Learning IoT in edge: Deep learning for the internet of things with edge computing. IEEE Network. 2018;**32**(1):96-101

[9] El-Sayed H, Sankar S, Prasad M, Puthal D, Gupta A, Mohanty M, et al. Edge of things: The big picture on the integration of edge, IoT and the cloud in a distributed computing environment. IEEE Access. 2017;**6**:1706-1717

[10] Sun X, Ansari N. EdgeIoT: Mobile edge computing for the internet of things. IEEE Communications Magazine. 2016;**54**(12):22-29

[11] Guilar NJ, Kleeburg TJ, Chen A, Yankelevich DR, Amirtharajah R. Integrated solar energy harvesting and storage. IEEE Transactions on Very Large Scale Integration (VLSI) Systems. 2009;**17**(5):627-637

[12] Giuppi F, NIoTaki K, Collado A, Georgiadis A. Challenges in energy harvesting techniques for autonomous self-powered wireless sensors. In: 2013 European Microwave Conference. Nuremberg, Germany: IEEE; 2013. pp. 854-857

[13] Yu H, Zhou J, Deng L, Wen Z. A vibration-based MEMS piezoelectric energy harvester and power conditioning circuit. Sensors. 2014;**14**(2):3323-3341

[14] Kanno I. Piezoelectric MEMS for energy harvesting. Journal of Physics: Conference Series. 2015;**660**(1):012001

[15] Gorlatova M, Sarik J, Grebla G, Cong M, Kymissis I, Zussman G. Movers and shakers: Kinetic energy harvesting for the internet of things. IEEE Journal on Selected Areas in Communications. 2015;**33**(8):1624-1639

[16] Caliò R, Rongala UB, Camboni D, Milazzo M, Stefanini C, de Petris G, et al. Piezoelectric energy harvesting solutions. Sensors. 2014;**14**(3):4755-4790

[17] Goudar V, Ren Z, Brochu P, Potkonjak M, Pei Q. Optimizing the

output of a human-powered energy harvesting system with miniaturization and integrated control. IEEE Sensors Journal. 2013;**14**(7):2084-2091

[18] Rao Y, McEachern KM, Arnold DP. A compact human-powered energy harvesting system. Energy Harvesting and Systems. 2014;**1**(1-2):89-100

[19] Kim S, Vyas R, Bito J, NIoTaki K, Collado A, Georgiadis A, et al. Ambient RF energy-harvesting technologies for self-sustainable standalone wireless sensor platforms. Proceedings of the IEEE. 2014;**102**(11):1649-1666

[20] Vullers RJ, van Schaijk R, Doms I, Van Hoof C, Mertens R. Micropower energy harvesting. Solid State Electronics. 2009;**53**(7):684-693

[21] Hassan HA, Nuaymi L, Pelov A. Renewable energy in cellular networks: A survey. In: 2013 IEEE Online Conference on Green Communications (OnlineGreenComm). IEEE; 2013. pp. 1-7

[22] Yang Y, Zhou K, Blaabjerg F. Enhancing the frequency adaptability of periodic current controllers with a fixed sampling rate for grid-connected power converters. IEEE Transactions on Power Electronics. 2015;**31**(10):7273-7285

[23] Available from: https://www.psma.com/HTML/newsletter/Q2_2012/page8.html

[24] Available from: https://www.powerelectronics.com/power-management/power-management-chapter-13-energy-harvesting

[25] Chiriac S, Rosales B. An ambient assisted living monitoring system for activity recognition–results from the first evaluation stages. In: Ambient Assisted Living. Berlin, Heidelberg: Springer; 2012. pp. 15-28

[26] Peter H, Raghu D. Energy Harvesting and Storage for Electronic Devices 2010-2020; 2010

[27] Adila AS, Husam A, Husi G. Towards the self-powered Internet of Things (IoT) by energy harvesting: Trends and technologies for green IoT. In: 2018 2nd International Symposium on Small-Scale Intelligent Manufacturing Systems (SIMS). Cavan, Ireland: IEEE; 2018. pp. 1-5